Mathematics:
Rhyme and Reason

MSRI Mathematical Circles Library

Mathematics: Rhyme and Reason

Mel Currie

MSRI
Mathematical Sciences Research Institute
Berkeley, California

AMS American Mathematical Society
Providence, Rhode Island

This volume is published with the generous support of the Simons Foundation and Tom Leighton and Bonnie Berger Leighton.

2010 *Mathematics Subject Classification.* Primary 00A09, 97A80.

For additional information and updates on this book, visit
www.ams.org/bookpages/mcl-22

Library of Congress Cataloging-in-Publication Data
Names: Currie, Mel (Melvin Robert), 1948- author.
Title: Mathematics : rhyme and reason / Mel Currie.
Description: Berkeley, California : MSRI Mathematical Sciences Research Institute ; Providence, Rhode Island : American Mathematical Society, [2018] | Series: MSRI mathematical circles library ; 22 | Includes bibliographical references and index.
Identifiers: LCCN 2018028165 | ISBN 9781470447960 (alk. paper)
Subjects: LCSH: Mathematics–Study and teaching (Higher) | Mathematics–Humor. | AMS: General – General and miscellaneous specific topics – Popularization of mathematics. msc | Mathematics education – General, mathematics and education – Popularization of mathematics. msc
Classification: LCC QA11.2 .C87 2018 | DDC 510.71/2–dc23
LC record available at https://lccn.loc.gov/2018028165

Dedicated to the memory of Angela E. Grant and Rudy L. Horne

Contents

Preface   ix

Chapter 1.   The Riddle   1

Chapter 2.   Primes   3

Chapter 3.   Some Geometry   9

Chapter 4.   Mysterious Pattern   17

Chapter 5.   Some Things Add Up. Some Don't.   21

Chapter 6.   A Tangential Remark   29

Chapter 7.   Plus or Minus   35

Chapter 8.   Making the Optimal Choice   41

Chapter 9.   Impossibilities   47

Chapter 10.   Magnitudes of Infinity   57

Chapter 11.   The Inevitable (Sperner's Lemma—The Brouwer
              Fixed-Point Theorem)   67

Chapter 12.   Consider the Sequence (Fibonacci and Golden Ratio)   81

Chapter 13.   What Are the Chances?   91

Chapter 14.   The Euler Line   99

Chapter 15.   The Dissertation   109

Chapter 16.   The Next Prime Number Is? (Gandhi's Formula)   113

Chapter 17.   Bulgarian Solitaire   119

Chapter 18.   Which is Bigger? (a^b versus b^a)   131

Chapter 19.   Fascinating   135

Chapter 20.   From the Sublime to the Ridiculous   141

Chapter 21.   A Few More Words   147

Photos and Pictures   151

Appendix A.   Notation, etc.   161

Appendix B.   Mysterious   167

Appendix C.   Impossibilities   171

Appendix D.   Magnitudes   173

Appendix E.   Fascinating   175

Preface

When I arrived on the Yale campus over fifty years ago, I was two months shy of my eighteenth birthday. I had never held a conversation with a mathematician and knew little to nothing about the culture of the mathematics community. But, somehow, I believed that I wanted to be a mathematician. Any savvy observer would have certainly concluded that my prospects were grim. Nonetheless, by hook *and* by crook, I eventually found a path.

I have fashioned a manuscript that I wish had been available to me when I was making my way through high school and dreaming of becoming a mathematician. Of course, I am trusting that such kids still exist. I suspect that many adults who feel that they missed the mathematics boat, and regret it, will find that this book resonates with them. I include what I deem to be mathematical gems and intersperse stories about mathematicians. The working title of this book was "Nursery Rhymes of Mathematics" because I hoped that the book would go a long way in establishing a cultural foundation in mathematics; an introduction to a perspective that most people are currently not exposed to sufficiently early in their lives. Grappling with many of the gems requires various levels of sophistication, but the tools needed are not beyond what many college-bound students will have mastered by twelfth grade. Most of the "nursery rhymes" require only basic high school algebra or geometry. The book is a kind of sampler. The chapters are arranged roughly in the order of difficulty and can be read independently of each other. Some students might take years to peruse all of them, with or without the support of a teacher.

I did not avoid humor in writing *Mathematics: Rhyme and Reason* and I was certainly willing to indulge my sentimental side in certain passages, since this book is also serving as something like a memoir. Topics covered include:

- The Infinitude of Primes
- Infinite Series
- No Four Points in the Plane with Pairwise Distances All Odd
- Magnitudes of Infinity/Existence of Transcendental Numbers
- Sperner's Lemma/Brouwer Fixed Point Theorem
- Binet's Formula

- The Euler Line
- Gandhi's Formula for the nth Prime
- Bulgarian Solitaire
- a^b versus b^a
- Sylvester's Problem

There are stories about Paul Erdős, Donald J. Newman, Abel, Yitang Zhang, Euler, J. Ernest Wilkins, Abraham Robinson, and others. Some of the stories are from firsthand knowledge. I believe, of course, that all of them are interesting.

I have always felt a greater kinship with philosophers and linguists than with physical scientists and engineers. Accordingly, the force of language in rigorous argumentation receives a somewhat greater emphasis than the clever use of visual display. I hasten to add that the obvious utility of the visual approach is not neglected.

It is my conviction that real headway in mathematics can only be made by following the aesthetic sense. Wonderful applications are often a byproduct of finding the poetry in the subject. There is a widely held belief that if we can just make it clear to young people how useful mathematics is, then their performance in the subject will improve. As opposed to other academic domains, there is generally no attempt to enhance what I will call an appreciation for mathematics and the culture in which it is embedded. At best, games and user-friendly applications are served up to make the confrontation more palatable. The emphasis is placed on mathematics as a tool, not on its aesthetic value. More often than not, the young person never encounters the humanity that lives in mathematics itself. The small gems and nuggets that are included in this thin volume represent, I hope, one small step toward addressing what I believe is a shortfall in the acculturation of young people in the field of mathematics. The reader will find, sprinkled among these mathematical nursery rhymes, stories about flesh-and-blood creatures, who call themselves mathematicians.

$$* * *$$

Thanks are due to Tatiana Shubin, who continually encouraged and prodded me as I wrote this book. My son-in-law Allen Welkie performed an invaluable service by setting up the LaTeX platform and getting the first draft encoded. My MSRI editor was Maia Averett, who offered insightful suggestions, especially regarding the chapter ordering. The AMS publisher Sergei Gelfand was always willing to go the extra mile in his support of the project. Thanks to Christine Thivierge who helped me navigate the shark-infested waters of *permissions*. Joe Moscati's rendering of my cartoons was perfect, paying homage to the primitive nature of my originals.

Of course, I thank my many readers. In no particular order, they were Rosemary Cordwell, Michael Parker, Joe Carmel, Arthur Denberg, Steve Kennedy, Steve Nordfjord, James Tanton, Lucille Moholt-Siebert, and

Ella Russell. I feel that I must single out Victor Marek and Stephanie Somer-sille, whose attention to detail combined with doggedness was nothing short of heroic.

Melvin Robert Currie
Baltimore, Maryland

Euclid alone has looked on Beauty bare

Edna St. Vincent Millay

Chapter 1

The Riddle

I grew up on the East End of Pittsburgh in what we called Little Italy. The Allegheny River was about a mile away, a walk due north along the railroad tracks through the woods. This walk was forbidden, but children on our dead-end street were largely unsupervised and some of us occasionally walked north. My mother and her siblings had also grown up on that street above the railroad tracks. It was a neighborhood that was largely white and Catholic. There was a smattering of African-American families that had been there for two generations. In the generation before mine these African-American families produced the pianist-composers Billy Strayhorn, Erroll Garner, Mary Lou Williams, and my uncle, Ahmad Jamal. The Billy Strayhorn that I knew lived a few blocks from me and was *the* Billy Strayhorn's nephew. I doubt that the topic of music ever came up in a conversation between Billy and me. I do know that I had a sense that perhaps I could make some contribution to society and that was due in part to being aware of my uncle's success. I do not know if Billy felt the same way.

We were children. Along with the other kids, my little brother Albert and I played constantly on the street and down the hill on the field next to the railroad tracks. The African-American adults on the street who were my mother's generation had a positive influence on me, especially the women. They were well educated and bookish. My mother's older brother had once dreamed of being an engineer, but had put those aspirations aside to make a living. The child who had the greatest impact on my mathematical development was five years older than me. Larry took Catholicism and mathematics seriously. He was passionate about games, especially chess, and was on his way to becoming a mathematics teacher. I happened to be in the neighborhood and a decent first pupil. In the eighth grade, when I was teaching myself first-year high school algebra, I would sometimes ask him questions about topics that I found in the book, but were beyond the scope of my course. However, the most important question was one he posed to me, a riddle.

When I think about this riddle now, I realize how simple it is. I also recognize that it captures an approach to precise argumentation that I found very satisfying, and I still do. Larry said, "Three men go into a dark room. On the table there are three green hats and two red hats. The three men all know this fact. They cannot see the color of the hats, but each puts one hat on. The remaining hats are removed and then the lights are turned on. The men are standing in single file so that man number 3 can see man number 2 and man number 1. Man number 2 can only see man number 1. Man number 1 can see no one. Man number 3 immediately says that he cannot deduce the color of his hat. Hearing this, man number 2 announces that he is also unable to deduce the color of his hat. Man number 1 then says, 'I know the color of my hat.' What is the color of man number 1's hat?"

Solving this riddle was a watershed event for me. My stepfather was somewhat peeved when I explained it to him and his friend. His friend confessed to not understanding my explanation but was laughing, amused by how fast I talked while producing the line of reasoning. His amusement seemed to annoy my stepfather even more.

There have been other moments like this for me. There was the time in high school when I did my first proof by contradiction. Many years later I looked up into the sky the night before my birthday and suddenly knew how to prove the theorem that became the crown jewel of my dissertation.

Chapter 2

Primes

When I was a senior in high school, I attended a lecture at the Carnegie Institute of Technology (now Carnegie Mellon University). The lecture was given by a mathematician from England, whose name I had probably forgotten before his presentation had even been completed. What I have never forgotten was being awakened to the mystery of mathematics. The speaker talked about open questions, some of which mathematicians had been trying to answer for centuries. That was fifty years ago. I was enthralled.

Although I cannot say for sure, it is likely that two of the open questions mentioned in that lecture were Fermat's Last Theorem and the Twin Prime Conjecture.

Fermat's Last Theorem is the claim that for all integers n, greater than or equal to 3, there are no *positive* integers x, y, z that satisfy:

$$x^n + y^n = z^n .$$

The legendary seventeenth century mathematician, Pierre de Fermat, claimed to have found a proof of this, but never published one. (Since just about all of the mathematicians in this book will be legendary, except for me, I will stop using this adjective.) Despite his stellar reputation, the mathematical world remained skeptical that Fermat had really proven it. More than three hundred years after Fermat's death, Andrew Wiles proved Fermat's Last Theorem. The year was 1994.

Let's move to the Twin Prime Conjecture, which remains unsettled. As a preliminary remark, we remind the reader that a prime number is an integer, greater than or equal to 2, which is only divisible by 1 and itself. The integers 2, 3, and 5 are examples of prime numbers. The integer 2 is, of course, the only even prime.

What is meant by "divisible"? One integer is divisible by another if there is no remainder when the division is completed. The integer 11 is not divisible by the integer 4 because there is a remainder of 3. The integer 39 is divisible by 13 because the remainder is zero. A fundamental truth in arithmetic is that every integer greater than 1 is either a prime itself or can be written in a unique way as the product of prime numbers. (We do not view a particular order of the factors as being different from any other order.) So, every integer greater than 1 is divisible by a prime.

The conjecture is that there are infinitely many pairs of primes that differ by 2, such as 3 and 5, 11 and 13, 17 and 19. Number theorists have been wrestling with this question for quite a long time. Somewhat surprisingly, the first known reference to it in the literature was in 1849 by de Polignac. Before we consider this conjecture, let's establish that there are infinitely many primes, our first "nursery rhyme." As was stated in the Preface, I view nursery rhymes as an important component of the cultural foundation. A proof that there are infinitely many prime numbers was produced 2400 years ago by the Greek mathematician Euclid (see the title of the Edna St. Vincent Millay poem at the end of the Preface). We must decide how to capture the infinitude of the collection of primes. Our agreement will be that if we can show that there must always be a bigger prime than the last one that we have found, that would prove the claim.

So Mary had a little prime, 2, and then she found 3, and then she found 5. The notion followed her to school one day and she had the temerity to tell the teacher that she had found three prime numbers, but was tired of looking for more. How could she be sure that there are others? I don't believe that I knew the answer to this when I was a senior in high school, even though I was taking calculus.

The teacher's response to Mary was to tell her that whenever she decided to stop, she could always be assured that there is one beyond the last one she has found. The teacher told her to take all the primes that she had found and imagine that she multiplied them all together: $2 \times 3 \times 5 \times \cdots \times p$, where the "$\cdots$" means "continue in this way" and p is the last prime that she has found. We will assume that she has not missed any primes less than p.

Now read the teacher's words in the same way that you read a poem for the first time. Savor and try to grasp each sentence before moving to the next one. Then read the whole "poem" several more times, until you have the significance in your bones.

We'll create an integer M. Our M will be your product plus 1:

$$M = (2 \times 3 \times 5 \times \cdots \times p) + 1$$

There are only two possibilities. M is either a prime number or it is not. If M is a prime, it is a prime bigger than any you have already found. If M is not a prime, consider the fact that it is divisible by a prime.

But it is not divisible by any prime that you have found so far, because dividing by any one of them always results in a remainder of 1.

So, M must be divisible by a prime not on your list and it must be bigger than the last one you found, since you have not missed any along the way.

So, in either of the two possible cases, there must be a bigger prime than the last prime you found.

It is useful to reflect on this proof. Instead of looking for a bigger one than the last one we have found, we could have agreed that if we can show that every finite set of primes is missing at least one prime, then the sequence of primes must be infinite. This statement is shown to be true by using the same line of reasoning that we have just used to show that there is always a prime bigger than the last one we have found. (Multiply all of the primes in your finite set together and add 1.) I like this way of establishing the infinitude of primes a bit better, but it is a matter of taste. We will be thinking about infinity again in this book, and it is good to recognize that there is more than one way to capture it and that, generally, being cagey is an asset.

Even though the Twin Prime Conjecture has not been proven, a major advance against the problem was made in 2013. I have a personal story that connects me in a tangential way to this event. That is to say, I managed to appear in a crowd scene.

Yitang Zhang was born in Shanghai. Around the age of nine, he came across a proof of the Pythagorean Theorem. He first learned of two major open questions, Fermat's Last Theorem and the Goldbach Conjecture, when he was ten. At age thirteen, during China's Cultural Revolution, he and his mother were sent to the countryside to work in the fields, where he languished for ten years. After the Cultural Revolution ended, Zhang entered Peking University at the age of twenty-three without having attended high school. He received a B.Sc. degree in 1982 and a M.Sc. degree in 1984. He then accepted a Fellowship at Purdue University in 1985, where he completed the Ph.D. degree in 1991.

After finishing the Ph.D., Zhang's story becomes bleak again. He was unable to establish himself in the academic world and struggled to make ends meet. It is said that for a while he lived in his car. After working at various jobs, a motel in Kentucky, and a Subway franchise, as well as doing delivery work for a New York City restaurant, Zhang landed a year-to-year position at the University of New Hampshire as a lecturer in 1999.

> "I was born for math," says Zhang. "For many years, the situation was not easy, but I didn't give up. I just kept going, kept pushing. Curiosity was of first-rank importance - - it is what makes mathematics an indispensable part of my life."[1]

At some point Zhang found his way to questions related to the Twin Prime Conjecture and spent three years working on them.

> On July 3, 2012, while visiting a friend's house in Colorado, he made his crucial breakthrough. "I tried to really make it a vacation. I didn't bring any book, notes, sheets, or my computer. I didn't use a pen," says Zhang. "But still I couldn't get rid of this completely. Sometimes I still tried to think about this point, this one small gap. How can we cross it?" As he was waiting to leave for a symphony concert that his friend was conducting, Zhang went into the backyard and started looking for some deer.

---

[1] *Yitang Zhang's Spectacular Mathematical Journey*, Kelly Devine Thomas, 2014.

"There are many deer sometimes'" says Zhang. "I didn't see any deer, but I got the idea."[2]

Eight months after not seeing any deer, Zhang put the finishing touches on the proof of what is called the Prime Gap Theorem. Specifically, he showed that there were infinitely many pairs of primes that differ by an integer k, for an integer k less than seventy million. Of course, the Twin Prime Conjecture states that this is true for $k = 2$ and the generalization is that this is true for every even, positive number, but until Zhang's breakthrough was achieved it was not known to hold for any k. No, we still don't know what the integer (or integers) k is (are). We just know by his theorem that there is an integer k less than seventy million that does the trick.

> "Never heard of him. Absolutely never heard of him," said Andrew Granville, a number theorist at the University of Montreal, in the movie *Counting From Infinity*. When Granville heard about the result and the techniques that Zhang used, he recalled saying, "There's no way that somebody I've never heard of has done this."[3]

In January, 2014, the annual Joint Mathematics Meetings were held at the Baltimore Convention Center. Over 6000 mathematicians from around the world attended, including Yitang Zhang and me. This was the year after Zhang's monumental achievement and I hoped to hear him talk about it. The meetings ran from January 15th through the 18th and Zhang was scheduled to talk on Thursday, the 16th. I attended his talk, but understood very little of it. Besides, he didn't really dwell on the research results that had thrust him into the limelight. He was moving on. I had noticed that Andrew Granville was scheduled to speak on Zhang's Prime Gap Theorem and the latest related developments the next day. I decided to attend.

Granville's talk was in a large room, which had the capacity to seat hundreds. I arrived early. Determined to find a seat that would allow me to bail out easily should I care to, I headed towards the chair in the very last row, right next to the aisle near the wall. That is to say, in the rear and as far as possible from the middle. As I moved toward the seat, there was a man walking ahead of me and he took the seat right in front of mine. I thought it was Zhang, which is a bit strange, since I was basing this only on the back of his head. After a few minutes, I walked forward to greet an acquaintance and on the way back to my seat established that it was indeed Zhang by looking at his conference badge.

By the time Granville launched into his talk, there was standing room only. He began with some remarks about Zhang's accomplishment, noting some of the hardship he had endured and the fact that the University of New Hampshire had given him a job and also afforded him time to work on research. Granville observed that when someone unknown has a major breakthrough of this magnitude, it is usually because they have taken an

---

[2] *Yitang Zhang's Spectacular Mathematical Journey*, Kelly Devine Thomas, 2014.
[3] www.quantamagazine.org/20150402-prime-proof-Zhang-interview/

entirely different approach. That was not true in this case. Zhang had used the very tools that the leaders in the field had developed. He refined the tools until he had sharpened them sufficiently to work his wonder. According to Granville, it was as if he had been working in this field for thirty years. Sitting in front of me, Yitang Zhang seemed to be just an interested observer. I doubt that anyone seated nearby even knew who he was. For about a half hour Granville discussed the technical underpinnings of the Prime Gap Theorem. I might describe my grasp of the matters at hand as tenuous, but it really was not nearly that good. At the break, Zhang stood to leave the room and I said, "Magnificent result." He smiled slightly and said, "Thank you."

At the end of the break Zhang returned and reclaimed his seat. Granville continued. I remained in a daze. As Granville wrapped up the second half of his talk, he returned to Zhang, the human being. His admiration was obvious. The Mathematical Sciences Research Institute in Berkeley was making a film about Zhang, he said. At this point I felt he was going to ask Zhang to stand. He did not. The talk ended and then Zhang stood. I looked at him and said, "Now you're going to be a movie star." He smiled and moved on.

Yitang Zhang was a co-recipient of the 2014 Frank Nelson Cole Prize in Number Theory. Although his work was done in obscurity, it was not done in a vacuum. One of Zhang's Cole Prize co-recipients was Daniel Goldston, whose response to being chosen for this honor appeared in the *Notices of the American Mathematical Society.*

Daniel Goldston:

> The mathematical work for which this prize has been awarded was aided by many mathematicians over many years, but I will not attempt to thank them all individually. Let me tell a story which starts in 1999, when, for many of us, the recent stunning work of Yitang Zhang would have seemed less likely than a proof of the Riemann Hypothesis. Cem [Yildirim] and I were both visiting MSRI in Berkeley, and one day he came in and said that rather than working on the paper we were supposed to be writing, he had been trying to work out a triple correlation divisor sum. We started working on this together and, after a few weeks, began to see that it was possible to work out asymptotic formulas for this type of sum. Over the rest of the term we continued to work on this, at first getting different answers each time we did the calculation but eventually tending toward only one answer. Cem went back to Turkey, but we continued our joint work by email, slowly working out asymptotic formulas for these sums. This was incremental research, the only kind I actually know how to do, where we used standard classical methods and neither knew nor expected any exciting applications. At the time with little kids

of ages 0, 2, and 4 in the house, and fragmented times for work, doing these calculations was ideal since they could be interrupted and then easily resumed. Finally, when the kids were 3, 5, and 7 in 2003, Cem and I thought we had made a breakthrough on gaps between primes but, while we received a lot of publicity, this did not help change the fact that the proof was wrong. Math can be a tough business, and while mathematicians often do not have much humility, we all have lots of experience with humiliation. Fortunately, in this case our work was not destined for the wastebasket, and in 2004 Green and Tao found a use for our formulas in their work on arithmetic progressions of primes, and in 2005, with János Pintz, we obtained the GPY method which proved new results on gaps between primes and provided part of the foundation for Zhang's great advance in 2013.[4]

For me the most amusing and poignant sentence is: "Math can be a tough business, and while mathematicians often do not have much humility, we all have lots of experience with humiliation." When the bough breaks the cradle will fall.... Every now and then you might be compensated for feeling inadequate; you find the mathematical equivalent of a gold nugget. You may love mathematics, but mathematics is never going to love you. If you can handle unrequited love, you'll be fine.

In addition to the Cole Prize, Zhang was awarded the 2013 Morningside Special Achievement Award in Mathematics, the 2013 Ostrowski Prize, and the 2014 Rolf Schock Prize in Mathematics. He was also a recipient of the 2014 MacArthur award, and was elected as an Academic Sinica Fellow during the same year.

---

[4] *Notices of the American Mathematical Society*, April 2014, p. 399.

Chapter 3

Some Geometry

It is known that human beings can have a change of heart. If the change is to the opposite point of view, we often say that they are voicing an opinion or belief that is 180 degrees from where they stood previously. Some people say that the change is 360 degrees. Those of us in the know point out that this means they have come full circle. They would be back where they started.

Taking a look at the rectangle $ABCD$, we imagine a person walking from A to B, then from B to C, from C to D, and then back to A to face B again.

The walker has made four turns, negotiating the angles at vertices A, B, C, and D. All of the angles are the same size and the walker has come full circle. We divide 360 degrees by 4 to get 90 degrees, and 90 is the value that we assign to each angle of a rectangle. We call an angle of 90 degrees a "right angle." Now we are ready to take a look at two geometric gems, one that almost every adult knows (hint) and one that few know.

We start with thinking about how we count. In general, counting is not a stumbling block. We can even use shortcuts. If we see that there are six rows, each with four chairs, it takes us no time to deduce that there are twenty-four chairs in the arrangement. We multiply instead of counting each chair individually. There is a smooth transition from counting objects in a rectangular array and computing the area of a rectangle that is, say, 6 feet by 4 feet. We see immediately that the rectangle contains twenty-four one-by-one squares. It is 24 square feet. So we move easily from counting to computing the area of a simple geometric object. It is child's play to come up with a formula for the area of a rectangle that is a units long and b units wide.

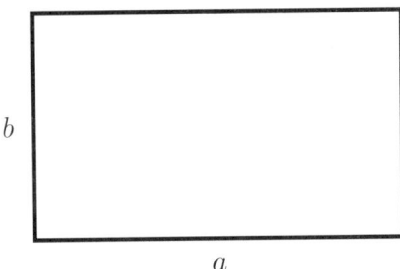

$$\text{Area} = ab$$

Well, we're making progress. We are getting there. Actually, I'm trying to get to the Pythagorean Theorem. So far we have made it all the way from counting to the area of rectangles. Let's cut the rectangle into two triangles of equal size:

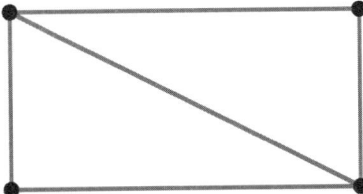

Since the two triangles that we have formed are the same (congruent), each triangle must have half the area of the rectangle that they cover:

$$\text{Area} = \frac{1}{2}ab.$$

Further, we see that each triangle contains a right angle, which added to the other two angles must be 180 degrees, since the angles of the two triangles cover the content of the angles in the rectangle.

Let's take four copies of one of the triangles and arrange them this way.

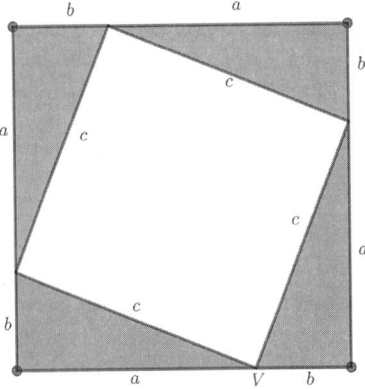

The inner white square has edges of length c. The angle at position V in the white region has to be 90 degrees, because the two adjacent angles sum to 90 degrees and the sum of all three angles must be 180 degrees. Of course, the same reasoning applies to the other angles in the white figure.

Note that the area of the white square is c^2 and remember that!

Now rotate the upper triangle on the left counterclockwise until its hypotenuse of length c aligns with the hypotenuse of the upper triangle on the right. Rotate the lower triangle on the left until its hypotenuse aligns with that of the lower triangle on the right. We have the picture below.

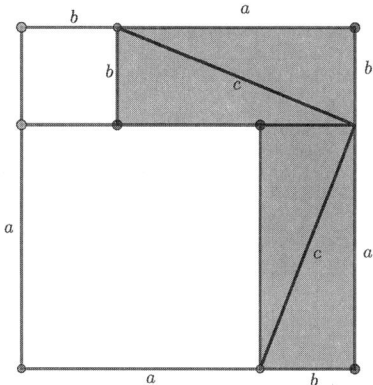

Note that the total area of the two white squares is $a^2 + b^2$. The area taken up by the gray triangles in the large square has not changed, so the remaining white space inside the large square must be the same as the white square before we rearranged the triangles. We have

$$a^2 + b^2 = c^2.$$

For those of you who prefer pulling out algebraic tools to moving the pieces in the puzzle around, here's another way to prove the theorem.

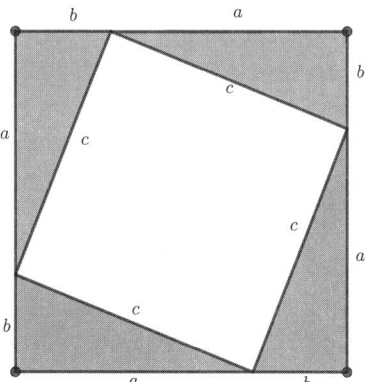

We have two squares. The larger square has edges of length $a + b$, so its area is

$$(a + b)^2 = a^2 + 2ab + b^2.$$

The area of the larger square is also equal to four times the area of a gray triangle plus the area of the white square:

$$4 * \frac{1}{2}ab + c^2 = 2ab + c^2.$$

We must then have

$$a^2 + 2ab + b^2 = 2ab + c^2.$$

Subtracting $2ab$ from both sides of the last equation again yields the *Pythagorean Theorem*:

$$a^2 + b^2 = c^2,$$

where c is the edge of the triangle opposite the right angle (the hypotenuse) and a and b are the other two edges. We were able to establish this by first counting and then seeing that counting told us something about area. The concept of area then told us something about length; ultimately distance.

* * *

Well, we had to work a bit, but we got there. Here's a nursery rhyme that might not be in your repertoire. Let's look at a quadrilateral. We just finished looking at rectangles, but now we place no conditions on the angles at the vertices, as long as the figure consists of four line segments, one from vertex A to vertex B, one from B to C, one from C to D, and finally a segment from D back to A again.

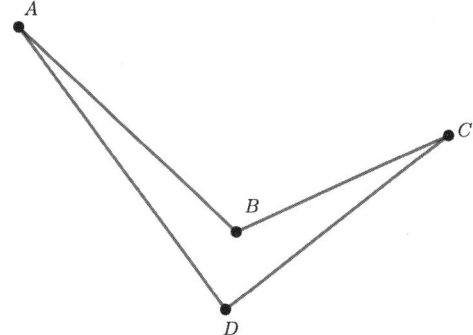

This quadrilateral looks a lot like a boomerang. Let's find the midpoints of each edge. I'll call the midpoint of edge AB, A', the midpoint of BC, B', etc. I claim that the quadrilateral $A'B'C'D'$ must be a parallelogram. That is, the new quadrilateral must have opposite sides parallel and equal in length. Take a look.

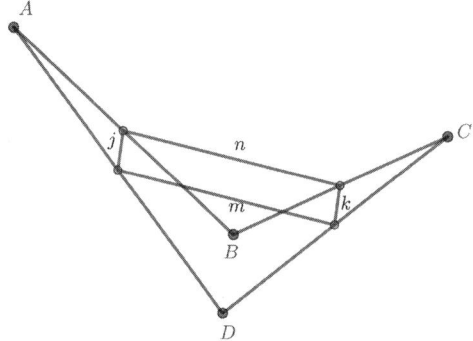

Our odd-looking quadrilateral has given rise to a parallelogram. Why must this always be so, no matter what quadrilateral we choose? Maybe we won't really be able to say why this is so, but we will be able to show that it *is* so.

We can show that this claim is true by returning to arithmetic. Consider the pathological case, where we draw all of the edges on the number line. For example, let $A = 2$, $B = 3$, $C = 5$, and $D = 13$. The midpoints of the edges are then $A' = 2.5$, $B' = 4$, $C' = 9$, and $D' = 7.5$. Opposite edges $A'B'$ and $C'D'$ have length 1.5 and opposite edges $B'C'$ and $D'A'$ have length 5. Clearly all of the edges are parallel, since they all lie on the number line.

As it turns out it is not much harder to establish this for points in the plane than it is on the number line.

For the reader who remembers the basics of similar triangles, the following proof will work just fine. We add two dotted lines to our graphic.

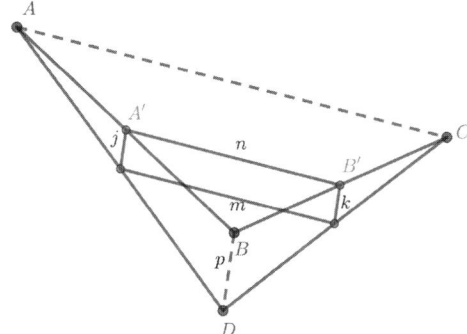

Segment n connects the midpoints of the edges AB and BC in triangle ABC. Since triangle ABC is similar to triangle $A'BB'$, segment n must be half the length of edge AC and parallel to it. The same argument holds for segment m in triangle ADC. It must be half the length of edge AC and parallel to it. We conclude that segments m and n are parallel to each other and equal in length. We can make the same observations using segment p to show that segments k and j are parallel and have the same length.

The proof we have just given is far removed from the simple arithmetic argument that we made in the case where all four points were on the number line. And we had to assume that the reader knows something about similar triangles. For the remainder of this section, I will try to bend things back to something that looks like, well, just arithmetic.

If the points are chosen in the plane, we can describe them using coordinates:

$$A = (a_1, a_2), B = (b_1, b_2), C = (c_1, c_2), D = (d_1, d_2).$$

The midpoints of AB, BC, CD, and DA are simply

$$A' = \left(\frac{a_1 + b_1}{2}, \frac{a_2 + b_2}{2} \right),$$

$$B' = \left(\frac{b_1 + c_1}{2}, \frac{b_2 + c_2}{2} \right),$$

$$C' = \left(\frac{c_1 + d_1}{2}, \frac{c_2 + d_2}{2} \right),$$

$$D' = \left(\frac{d_1 + a_1}{2}, \frac{d_2 + a_2}{2} \right),$$

respectively.

Let's see that this is true for A'. The slope of the line containing A and B is

$$\frac{a_2 - b_2}{a_1 - b_1}.$$

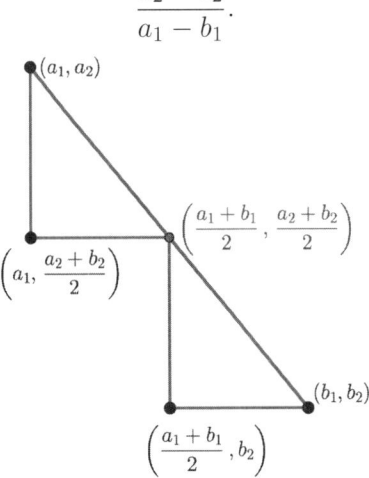

A quick calculation shows that the slope of the line containing A and A' is also

$$\frac{a_2 - b_2}{a_1 - b_1}.$$

So A' is on the same line as the line passing through the points A and B.

There is the question of equidistance. How do we compute distances between points in the Cartesian plane? Well, we use the Pythagorean Theorem, which we have just derived. Again, we'll stick with A', understanding that the other three cases are completely analogous.

A' is equidistant from A and B, the distance in both cases being

$$\sqrt{\left(\frac{a_1 - b_1}{2} \right)^2 + \left(\frac{a_2 - b_2}{2} \right)^2}.$$

All that we have left to do is show that the line segments $A'B'$ and $C'D'$ are parallel and have equal length, and the same for $B'C'$ and $D'A'$. That's just more of the kind of exercise that we have already carried out.

The distances from

$$A' = \left(\frac{a_1 + b_1}{2}, \frac{a_2 + b_2}{2} \right) \text{ to } B' = \left(\frac{b_1 + c_1}{2}, \frac{b_2 + c_2}{2} \right) \text{ and from}$$

$$C' = \left(\frac{c_1 + d_1}{2}, \frac{c_2 + d_2}{2} \right) \text{ to } D' = \left(\frac{d_1 + a_1}{2}, \frac{d_2 + a_2}{2} \right) \text{ are both}$$

$$\sqrt{ \left(\frac{a_1 - c_1}{2} \right)^2 + \left(\frac{a_2 - c_2}{2} \right)^2 }.$$

The slopes of the lines through the corresponding vertices are both

$$\frac{a_2 - c_2}{a_1 - c_1}.$$

The other two edges are also obedient!

What we have done moves us very close to vector arithmetic. (See Appendix A.) We could have simply added multiples of our points coordinate-wise. The midpoints of the edges AB, BC, CD, and DA are

$$A' = \frac{1}{2}(A + B) = \left(\frac{a_1 + b_1}{2}, \frac{a_2 + b_2}{2} \right),$$

$$B' = \frac{1}{2}(B + C),$$

$$C' = \frac{1}{2}(C + D),$$

$$D' = \frac{1}{2}(D + A).$$

We then see that opposite edges have the same length and are parallel by subtracting:

$$A' - B' = \frac{1}{2}(A - C),$$

$$B' - C' = \frac{1}{2}(B - D),$$

$$D' - C' = \frac{1}{2}(A - C),$$

$$A' - D' = \frac{1}{2}(B - D).$$

These differences produce vectors that point in the same direction and have the same length as the edges of the quadrilateral $A'B'C'D'$. Using vectors simplifies everything and takes us back to performing the same arithmetic that we carried out when our points were all on the number line.

Many of the lovely truths of plane geometry can easily be explained using vector arithmetic. On the other hand, attempts to establish them using geometric tools can sometimes be arduous, even though it was not in this case. One caution: you do have to make a convincing argument that the quadrilateral you've drawn covers all the cases. My colleague and friend, Nathaniel Dean, wandered into a minefield when he tried to prove our interesting fact about quadrilaterals and got wrapped around the axle. It took him a long time, because he felt he had to consider many cases. When I pointed out the error of his ways (just use vectors!), he exclaimed, disgusted with himself, "You're right. It's trivial." Translating this into laymen's terms it means, "Why am I such an abject failure in life?" In fact, if another mathematician says to me, "Mel, what you have done is non-trivial," the translation is actually something like a compliment.

Another advantage of using vectors is that we see immediately that there is no reason that we have to require the four vertices to lie in the same plane. They can be in three dimensions, four dimensions, ... The vector argument is free of dimension restrictions. The edges can even cross each other!

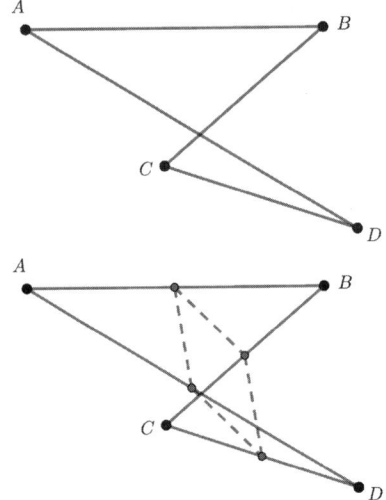

We could draw our first edge from Pittsburgh to Cincinnati, then make our second edge one inch long in some arbitrarily chosen direction, then draw our third edge to Fairbanks and our fourth edge back to Pittsburgh. Connecting the midpoints, we would have a parallelogram. Of course, sections of the edges might be underground or above the ground. As my friend Jim Schatz once said about this little fact, it demonstrates that order can frequently be found where there is apparently only chaos.

Chapter 4

Mysterious Pattern

In my mathematical youth I discovered a sequence that at the time I found exciting. In fact, I still do. Over the years I showed it to a number of people, because I wanted to know if anyone had ever seen it. No one that I asked had seen it. Years went by and I stopped asking.

$$a_0 = 2,$$

$$a_1 = 2\sqrt{2},$$

$$a_2 = 2^2\sqrt{2 - \sqrt{2}},$$

$$a_3 = 2^3\sqrt{2 - \sqrt{2 + \sqrt{2}}},$$

$$a_4 = 2^4\sqrt{2 - \sqrt{2 + \sqrt{2 + \sqrt{2}}}},$$

$$a_5 = 2^5\sqrt{2 - \sqrt{2 + \sqrt{2 + \sqrt{2 + \sqrt{2}}}}},$$

$$a_6 = 2^6\sqrt{2 - \sqrt{2 + \sqrt{2 + \sqrt{2 + \sqrt{2 + \sqrt{2}}}}}},$$

. . . and so on.

So what's so nice about this sequence beyond the pattern? Well,

$$a_0 = 2,$$

$$a_1 = 2\sqrt{2} = 2.828427124746\ldots,$$

$$a_2 = 2^2\sqrt{2 - \sqrt{2}} = 3.061467458920\ldots,$$

$$a_3 = 2^3\sqrt{2 - \sqrt{2 + \sqrt{2}}} = 3.121445152258\ldots,$$

$$a_4 = 2^4\sqrt{2 - \sqrt{2 + \sqrt{2 + \sqrt{2}}}} = 3.136548490545\ldots,$$

$$a_5 = 2^5\sqrt{2 - \sqrt{2 + \sqrt{2 + \sqrt{2 + \sqrt{2}}}}} = 3.140331156954\ldots,$$

$$a_6 = 2^6\sqrt{2 - \sqrt{2 + \sqrt{2 + \sqrt{2 + \sqrt{2 + \sqrt{2}}}}}} = 3.141277250932\ldots.$$

At this point there's a good chance you have guessed that this sequence is getting closer and closer (converges) to $\pi = 3.1415926\ldots$. But why is this so?

If you inscribe a square in a circle of radius 1, you see that by using the Pythagorean theorem, which we proved in the previous chapter, each edge must have length $\sqrt{2}$. It follows that its area is 2, which is the value of a_0. A square is an example of a regular (edges of equal length, angles of equal measure) convex polygon.

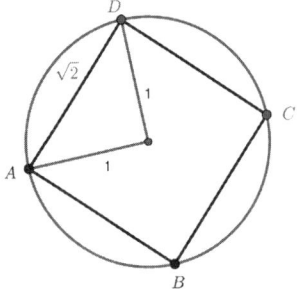

Below is a picture of a regular octagon inscribed in a circle of radius 1. Its area corresponds to a_1.

Our sequence corresponds to the areas of regular convex polygons with 2^n edges. (A detailed derivation is provided in Appendix B.) It is not surprising that as we look at an increasing number of edges the areas of the polygons converge to π, which is the area of a circle of radius 1. This is essentially Archimedes' Method of Exhaustion, which he used over two thousand years ago. What continues to captivate me is the way the sequence looks.

A few years ago I again tried to find out if this sequence was known. By then, I had the convenience of the internet. I found that the sequence was known in a somewhat different form and that this form was discovered by the mathematician Francois Viète about five hundred years before I discovered mine. So I was a bit late. Viète wrote it as an infinite product and I'll display the first few factors:

$$\frac{2}{\pi} = \frac{\sqrt{2}}{2} \times \frac{\sqrt{2+\sqrt{2}}}{2} \times \frac{\sqrt{2+\sqrt{2+\sqrt{2}}}}{2} \times \ldots.$$

A casual glance at this formula and the one I came up with probably does not reveal how closely related they are. Looking at a_3 of my sequence and comparing it with Viète's formulation illustrates what is going on. The reciprocal of the first three factors of Viète's formula multiplied by 2 is

$$\frac{2^4}{\sqrt{2+\sqrt{2+\sqrt{2}}} \times \sqrt{2+\sqrt{2}} \times \sqrt{2}} = \frac{2^3\sqrt{2}}{\sqrt{2+\sqrt{2+\sqrt{2}}} \times \sqrt{2+\sqrt{2}}}.$$

We had to multiply by 2, because the reciprocal of Viète's formula gives us an expression for $\frac{\pi}{2}$, not for π. On the other hand, if we consider a_3, we have

$$a_3 = 2^3\sqrt{2 - \sqrt{2+\sqrt{2}}} = \frac{2^3\sqrt{2 - \sqrt{2+\sqrt{2}}}}{1} \times \frac{\sqrt{2+\sqrt{2+\sqrt{2}}}}{\sqrt{2+\sqrt{2+\sqrt{2}}}}$$

$$= \frac{2^3\sqrt{2 - \sqrt{2}}}{\sqrt{2+\sqrt{2+\sqrt{2}}}} = \frac{2^3\sqrt{2 - \sqrt{2}}}{\sqrt{2+\sqrt{2+\sqrt{2}}}} \times \frac{\sqrt{2+\sqrt{2}}}{\sqrt{2+\sqrt{2}}}$$

$$= \frac{2^3\sqrt{2}}{\sqrt{2+\sqrt{2+\sqrt{2}}} \times \sqrt{2+\sqrt{2}}}.$$

There you have it. I do wonder about connections to the past. This sequence connects me to a man born five centuries before I was, Viète. There is another connection to that past that I think about often. It is to my mother's paternal grandfather, who was born in 1856 and named after the man elected President the month he was born, James Buchanan Jones. I was born ninety-two years later on the day Truman upset Dewey, but was spared being named after the winner. In early 1865, as an eight-year-old slave, my great grandfather James saw the Union troops sweep through Lunenburg County, Virginia, perhaps on their way to Appomattox to accept

Lee's surrender. By the time he was an adult he had acquired a tobacco farm and was raising a large family. Although he received no formal schooling, he somehow learned to read and write. On occasion, when he had a few minutes to himself, he would open a worn book, an algebra textbook. This last bit of oral history was not conveyed to me until I was an adult and already on my own mathematical path. James Buchanan's son, my grandfather Robert Jones, came to Pittsburgh right after World War I and was an open-hearth steel worker. My paternal grandfather had a wallpapering business. My maternal grandparents came from Virginia and my paternal grandparents from North Carolina. It was the beginning of the Great Migration. They all eventually settled on the East End, where there were clusters of African-American families embedded in a largely Italian and Catholic neighborhood. My mother and her siblings made it through the Great Depression. That's two "Greats" in this paragraph. As I write this, I fear that a third "Great" may be waiting in the wings.

I can find no documented genealogical path to Viète. The Mathematics Genealogy Project allows me to track my doctoral line to giants such as Bernoulli and Euler, but the lines become very faint in the 1500s, when Viète lived. Nonetheless, I will assert a kinship.

As I mentioned earlier in the chapter, a derivation of my sequence is in Appendix B. Of course, the reader may want to take a crack at deriving it before peeking.

Chapter 5

Some Things Add Up. Some Don't.

When our daughter was in fourth or fifth grade, I asked her what she thought about the sum

$$\frac{1}{2} + \frac{1}{4} + \frac{1}{8} + \ldots.$$

Before she tried to answer, I told her that she could give me her thoughts later, perhaps the next day or the day after. A few days later at dinner I asked her if she had thought about my question. Her answer was, "Not much." I said that I would be glad to hear what her thoughts were, even if not fully developed. She said, "I think it adds up to 1."

Petra's answer was correct. She had surprised me. I asked her how she came to this conclusion. This was her explanation: If you start off with a piece of paper that is one unit long and cut it in half, you have two half units. If you take one of them and cut it in half, you produce two one-quarter units, etc. All the pieces together are still one unit long.

Her answer would have been a fitting response to the Greek philosopher, Zeno, who used this very sum to build his famous paradox, Zeno's Paradox. Zeno used the above sum to show that motion was impossible. To get from A to B one must first go one half of the distance to B, then one half of what is left, then one half of the remaining distance, and so forth,

It is puzzling to me how this paradox ever became famous. There were a lot of smart people in ancient Greece. Why wasn't it just brushed aside? Well, tackling this philosophical issue would take us too far afield. We will just stick with the little girl's explanation. Effectively, she said that we can always imagine breaking a finite object into infinitely many parts. Of course, if we were to take the same amount of time to make each break we would never get done.

The 19th century mathematician Charles Dodgson, better known as Lewis Carroll, played around in his book *Through the Looking Glass*, with a Zeno-paradox-like description of motion:

"Well, in our country," said Alice, still panting a little, "you'd generally get to somewhere else - if you run very fast for a long time, as we've been doing."

"A slow sort of country!" said the Queen. "Now, here, you see, it takes all the running you can do, to keep in the same place.

If you want to get somewhere else, you must run at least twice as
fast as that!"

Zeno has left his mark not just on mathematics. Some accounts of his
life indicate that he was tortured after his participation in a failed attempt
to overthrow a monarch and bit off either the ear or the nose of said torturer,
perhaps the monarch himself, before he died.

It is interesting to note that most people readily accept the equality

$$\frac{1}{3} = 0.3333\ldots,$$

but many people struggle when we write

$$\frac{1}{3} = \frac{3}{10} + \frac{3}{100} + \frac{3}{1000} + \frac{3}{10000} + \cdots.$$

What we have on the right-hand side of the equality is called a geometric
series and that is really what $0.333\ldots$ is too. In fact, it's the same geometric
series.

For a discussion of summation and limit notation, which we are about
to use, see Appendix A. An infinite geometric series is a sum of the form

$$a + ar + ar^2 + ar^3 + \cdots = \sum_{n=0}^{\infty} ar^n.$$

In the case of our series for $\frac{1}{3}$, $a = \frac{3}{10}$ and $r = \frac{1}{10}$.

If we just consider the first n terms of the series, we have

$$a + ar + ar^2 + ar^3 + \cdots + ar^{n-1} = a\left(1 + r + r^2 + r^3 + \cdots + r^{n-1}\right)$$

$$= a\frac{1 - r^n}{1 - r}$$

$$= \frac{a}{1 - r} - \frac{ar^n}{1 - r},$$

since

$$\left(1 + r + r^2 + r^3 + \cdots + r^{n-1}\right)(1 - r) = 1 - r^n.$$

If $-1 < r < 1$, then

$$\lim_{n \to \infty} \frac{a}{1 - r} - \frac{ar^n}{1 - r} = \frac{a}{1 - r}.$$

Given the restrictions that we placed on r, we can get r^n as close to zero
as we like by simply making n sufficiently large. Making n large forces the
second term on the left side of the equality to zero as well.

We conclude that

$$\sum_{n=0}^{\infty} ar^n = \frac{a}{1 - r}.$$

We see that if
$$a = \frac{3}{10} \text{ and } r = \frac{1}{10},$$
$$\frac{a}{1-r} = \frac{1}{3}.$$

As we said above, the restriction on the use of the $\frac{a}{1-r}$ formula for the sum of an infinite geometric series is that r has to be less than 1 and greater than -1. If r is outside that range, r^n gets large without bound and our series does not "add up." It is not summable. But the formula for the sum of the first n terms holds without any restriction on r:

$$\sum_{k=0}^{n-1} ar^k = a\frac{1-r^n}{1-r}.$$

* * *

There is a charming story about Carl Friedrich Gauss. (Charming stories always have many versions.) One day, in grade school, Gauss was misbehaving. His teacher gave him an assignment intended to occupy his time and settle him down. His teacher asked him to add up the numbers from one to one hundred. As soon as the teacher's back was turned, only seconds later, young Carl started his shenanigans again. The teacher asked him why he was not working on his assignment. The response: "I'm done. The total is 5050." What had Carl done? He had simply noticed that

$$1 + 100 = 101,$$
$$2 + 99 = 101,$$
$$3 + 98 = 101$$
$$\vdots$$
$$50 + 51 = 101.$$

There are fifty pairs that add to 101, and 50 times 101 is 5050. Let the games begin!

Gauss had derived the formula

$$1 + 2 + 3 + 4 + \cdots + n = \sum_{k=1}^{n} k = \frac{n(n+1)}{2}.$$

Numbers of this form are called *triangular* numbers. For example, the fourth triangular number is 10, corresponding to $n = 4$. We will encounter triangular numbers again in the *Bulgarian Solitaire* chapter.

Here's a hint on the origin of the term triangular. Ten dots.

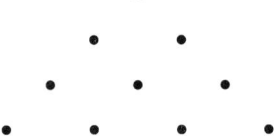

In the Gauss story, the $(n+1)$ in the numerator is 101 and the $n/2$ factor represents the $50 = 100/2$ pairs that add up to 101. Another way to look at this formula is to view $(n+1)/2$ as the average value of the sequence of integers from one to one hundred. Multiplying the average value of a set by the number of values in the set, n, gives us the total. Computing the average in this case is easy because the numbers are evenly spaced and we know the largest and the smallest value.

If Gauss's teacher had asked Gauss to add up the squares of the integers from one to one hundred, he might have been able to keep him settled down a bit longer:

$$1^2 + 2^2 + 3^2 + 4^2 + \cdots + n^2 = \sum_{k=1}^{n} k^2 = \frac{n(n+1)(2n+1)}{2}.$$

How can we see that this is true? We could prove that it is true by induction, but how would we even suspect that the formula we have above is a candidate? (Proof by induction is covered in Appendix A.)

Here is a clever way to derive it directly. Notice that

$$\sum_{k=1}^{n} k^3 - (k-1)^3 = \left(1^3 - 0^3\right) + \left(2^3 - 1^3\right) + \left(3^3 - 2^3\right) + \cdots$$
$$+ \left[(n-1)^3 - (n-2)^3\right] + \left[n^3 - (n-1)^3\right] = n^3.$$

We simply noted that all the terms cancel except for n^3. It is also true that

$$\sum_{k=1}^{n} k^3 - (k-1)^3 = \sum_{k=1}^{n} \left(3k^2 - 3k + 1\right) = 3\left(\sum_{k=1}^{n} k^2\right) - 3\left(\sum_{k=1}^{n} k\right) + n.$$

So

$$3\left(\sum_{k=1}^{n} k^2\right) - 3\left(\sum_{k=1}^{n} k\right) + n = n^3.$$

Gauss has already shown us that the second summation on the right is $n(n+1)/2$. Using that and a little manipulation we have

$$\sum_{k=1}^{n} k^2 = \frac{n^3}{3} + \frac{n(n+1)}{2} - \frac{n}{3} = \frac{n(n+1)(2n+1)}{6}.$$

I do not believe that Gauss would have gotten this in grade school. Take a second to see what this means about the average of the squares of the first n integers. It's

$$\frac{(n+1)(2n+1)}{6}.$$

$$* * *$$

Returning to infinite sums, consider

$$1 + \frac{1}{2} + \frac{1}{3} + \frac{1}{4} + \frac{1}{5} + \cdots.$$

This series is called the *harmonic* series because of its relationship to the tones produced by a vibrating string. What are we to make of this sum? Have we just broken up a finite entity into infinitely many pieces again? Consider the following:

$$1 > \frac{1}{2},$$

$$1 + \frac{1}{2} > \frac{1}{2} + \frac{1}{2},$$

$$1 + \frac{1}{2} + \left(\frac{1}{3} + \frac{1}{4}\right) > \frac{1}{2} + \frac{1}{2} + 2\left(\frac{1}{4}\right) = \frac{1}{2} + \frac{1}{2} + \frac{1}{2},$$

$$1 + \frac{1}{2} + \frac{1}{3} + \frac{1}{4} + \left(\frac{1}{5} + \frac{1}{6} + \frac{1}{7} + \frac{1}{8}\right) > \frac{1}{2} + \frac{1}{2} + \frac{1}{2} + 4\left(\frac{1}{8}\right) = \frac{1}{2} + \frac{1}{2} + \frac{1}{2} + \frac{1}{2}.$$

We are beginning to see the following pattern:

The first term is greater than one half. The sum of the first two terms is greater than two times one half. The sum of the first four terms is greater than three times one half. In general, the sum of the first 2^n terms is bigger than $\frac{n+1}{2}$. If each of these terms represented a distance, all of them taken together would certainly be larger than any number we might choose to name. We could never get to the end of the route. The terms do not represent the division of a finite length into infinitely many pieces. Unlike our "Zeno" sum, this series is not summable.

$$* * *$$

Sometimes it is not just that things add up. It is what they add up to. If we add up the digits in the number 294, we see that the sum is divisible by 3:

$$294 = 2(100) + 9(10) + 4 = 2(99) + 9(9) + (2 + 9 + 4) = 2(99) + 9(9) + 15.$$

Every term in the rightmost expression is divisible by 3, therefore 294 must be.

Generally, the three-digit number

$$xyz = 100x + 10y + z = 99x + 9y + x + y + z.$$

If $x + y + z$ is divisible by 3, then xyz must be. It is clear that this holds for any number, regardless of how many digits it has. We can make a completely analogous statement about divisibility by 9.

There is another goodie. It's divisibility by 11. Keep in mind that 9 and 11 straddle 10. This is, in fact, the reason we can come up with quick ways to detect whether or not an integer, written in base 10, is divisible by 9 or 11. Generally, the three-digit number

$$xyz = 100x + 10y + z$$

is divisible by 11, if $x - y + z$ is.

We can see this by writing

$$xyz = 100x + 10y + z = 99x + 11y + x - y + z.$$

Clearly, if $x - y + z$ is divisible by 11 then xyz is.

What about

$$xyza = 1000x + 100y + 10z + a = 1001x + 99y + 11z - (x - y + z - a)?$$

Seeing that 1001, 99, and 11 are all divisible by 11, our claim for three digits seems to extend in a natural way to four digits. There is a key to showing generally that if

$$\sum_{k=0}^{n} (-1)^k x_k$$

is divisible by 11, then

$$\sum_{k=0}^{n} 10^k x_k \text{ is too.}$$

The key is to show that the expressions

$$10^{2n} - 1 \text{ and } 10^{2n+1} + 1$$

are divisible by 11 for $n \geq 0$.

We can prove this by induction for the case of the even exponent and then recognize quickly that the odd exponent case follows the same way. (The odd exponent case is proven in Appendix A.)

Basis for induction: For $n = 0$, $10^{2n} - 1 = 1 - 1 = 0$, which is obviously divisible by 11.

We now show that if it is true for $n = k$, it is true for $n = k + 1$:

$$10^{2(k+1)} - 1 = 10^2 \left(10^{2k}\right) - 1 = 10^2 \left(10^{2k} - 1\right) + \left(10^2 - 1\right).$$

In the rightmost expression the first term is divisible by 11 by the induction hypothesis and the second term is 99. That completes the induction. Remarking that

$$10^{2n} x = \left(10^{2n} - 1\right) x + x \text{ and } 10^{2n+1} x = \left(10^{2n+1} + 1\right) x - x$$

should allow the reader to put this matter to rest.

I sometimes allow myself to become distracted in traffic. I check the divisibility by 9 of license plate numbers. (For some reason, I never find myself doing it for 11.) It is not enough to check the divisibility by 9 of the number on the plate. I also want to know if the quotient is divisible by 3 or 9. Things can pile up. I once happened upon $6561 = 3^8 = 9^4$ in traffic.

Chapter 6

A Tangential Remark

One of the central ideas in the calculus is *rate of change*. If there is a relationship between a quantity y and another quantity x, we are often interested in how y is changing as x changes. A classic example is looking at the relationship between distance traveled and elapsed time; velocity. If the relationship between distance and time does not change, the graph is simply a straight line and the rate of change is in that case constant. For example, if the object is moving at a constant rate of 15 feet per second, the rate of change would be, in fact, the slope of the line that describes the relationship. In this case, if x represents time and y represents distance, the slope would be 15. If the relationship does change, more sophistication is required to compute the rate of change at any given time. In general, we would need the differential calculus.

I propose that we do a little faith-based mathematics as a substitute for the differential calculus. (This is just entertainment.) Calculus tells us that if we graph the relationship between x and y, the slope of the tangent line at any given point of the curve is the value of the rate of change of y as x changes. We will take that as an article of faith. We want to know if we can find this tangent line without the power of calculus. Do we really need the work of Leibniz and Newton to find the slope of a tangent line?

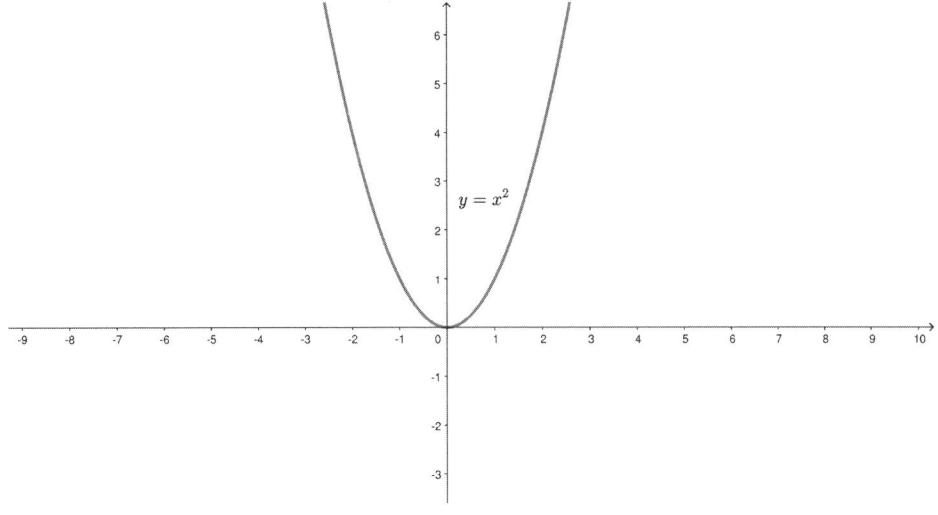

We have to figure out what a tangent line is. Do we know one when we see it? Pictured above is the graph of a relationship between y and x, where y corresponds to the vertical axis and x to the horizontal one.

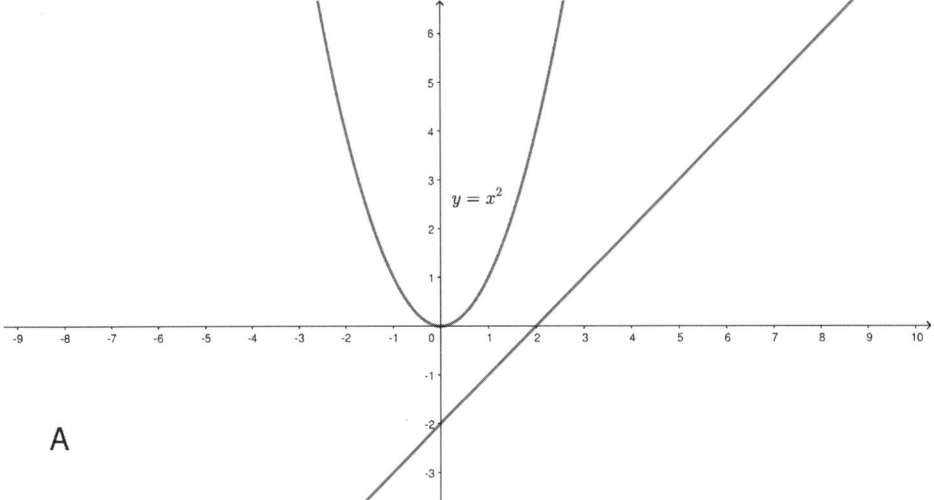

A

I am not willing to accept that the line pictured in graphic A is tangent to the curve, because my faith in my understanding of language tells me that the line has to touch the curve to be tangent to it. Try again.

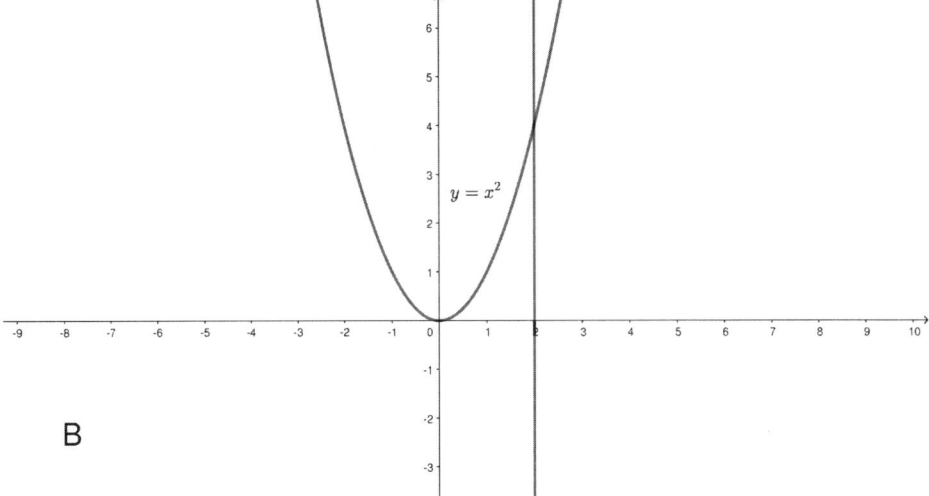

B

No! Graphic B is an affront! Let's exclude all vertical lines as possibilities.

How about the picture below in graphic C?

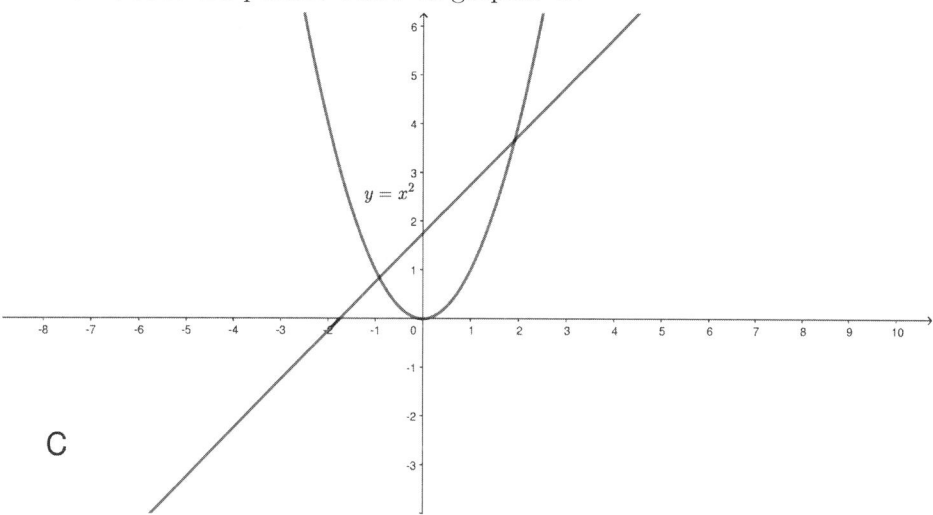

C

This is another candidate to reject. A tangent line must "kiss" the curve at the point of tangency, as the line in D illustrates.

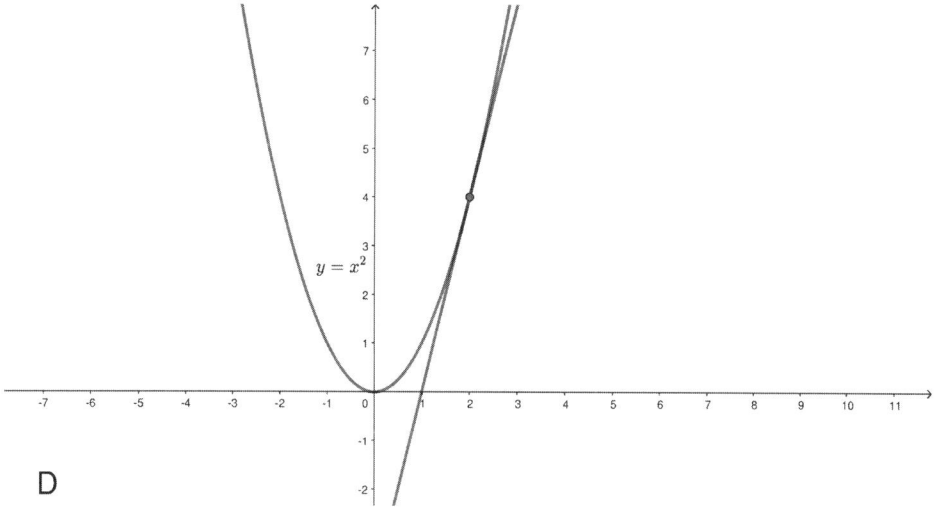

D

So, for this particular curve, we have rejected all vertical lines, lines that intersect this curve at more than one point, and lines that do not intersect the curve at all. We are only going to allow lines that intersect the curve at one point and only one point.

Since we have ruled out vertical lines, all of the lines still under consideration have the form

$$y = mx + \beta,$$

where m and β are constant. So we want to find x and y that satisfy both

$$y = mx + \beta$$

and
$$y = x^2.$$
If we do find a pair x and y that satisfy both equations, it must be true that
$$x^2 = mx + \beta \text{ or } x^2 - mx - \beta = 0.$$

It might seem that we are in a downward spiral, but we will keep at it nonetheless. There is a nice little formula, which undoubtedly many readers have seen, for finding solutions to equations like the one we are considering. It is the quadratic formula, and we will derive it now so that we can ease our way forward.

Recall that

(*) $$(u + w)^2 = u^2 + 2uw + w^2.$$

We are considering
$$ax^2 + bx + c = 0.$$

We derive the quadratic formula by a technique called *completing the square*. We would like to take advantage of the identity (*) by casting the terms of the quadratic equation in the roles of u and w. To facilitate this we multiply both sides of our equation by $4a$ to produce
$$4a^2x^2 + 4abx + 4ac = 0.$$

If we cast $4a^2x^2 = (2ax)^2$ in the role of u^2 and b in the role of w, we need to throw in b^2 to "complete" the square. We may add b^2 to the left side of the equation as long as we maintain the equality by adding it to the right side also. We then have
$$(2ax)^2 + 2ax\,(2b) + b^2 + 4ac = b^2,$$
$$(2ax)^2 + 2ax\,(2b) + b^2 = b^2 - 4ac,$$
$$(2ax + b)^2 = b^2 - 4ac,$$
$$2ax + b = \pm\sqrt{b^2 - 4ac},$$
$$x = \frac{-b \pm \sqrt{b^2 - 4ac}}{2a}.$$
The last equality is what we call the quadratic formula.

We were looking to solve
$$x^2 - mx - \beta = 0$$
for x. We have in our case $a = 1$, $b = -m$, and $c = \beta$. For what it's worth, we get
$$x = \frac{m \pm \sqrt{m^2 + 4\beta}}{2}.$$

We recognize that if the expression under the square root sign is anything but zero, our line will intersect the curve at more than one place or not at all.

We have said that such lines cannot be tangent to the curve. So, we must have

$$m^2 + 4\beta = 0, \text{ and thus } x = \frac{m}{2}.$$

It follows that the tangent line to the curve at the point (x, x^2) has slope

$$m = 2x.$$

Lo and behold, this is exactly what the calculus tells us about this curve, this *parabola*. The rate of change of y with respect to x is $2x$. Surprisingly, we did not use any of the machinery of calculus. We just waved our hands and moved ahead with faith in our ability to size up the situation. This story is another result of my youthful explorations. I will insist that it is not simply a bogus line of reasoning that happens to lead to something that is true. An example of specious "reasoning" leading to truth is demonstrated in this sleight of hand, cancelling the sixes:

$$\frac{2\cancel{6}}{\cancel{6}5} = \frac{2}{5}.$$

What we really have presented with our tangent lines is an ad hoc approach with a compelling, perhaps even pleasing, argument that unfortunately will not take us much further. We should ask ourselves what other polynomials we can handle in this way. Generally, tangent lines for cubic polynomials intersect the curve in more than one place. Further, our parabola involved a polynomial of degree two and we exploited the quadratic formula (a degree-two device). Is there a cubic formula, a quartic formula, or a quintic formula? The answers are yes, yes, and no, respectively. Trying to extend this approach to finding tangent lines is doomed. The wheels would come off.

Slowing down, the reader is perhaps puzzled by the "no" answer in the preceding paragraph. That was to say that there is no quintic formula that can be written simply in terms of the standard operations: multiplication/division, addition/subtraction, and root taking. It has been proven that no such general formula exists. We will discuss this in the *Impossibilities* chapter.

Chapter 7

Plus or Minus

In the next chapter, I write about finding the optimal choice, which in the world of mathematics is sometimes possible. In our everyday dealings, that is almost never the case. Often, we do not even have a choice. I did not get to choose my teachers. Looking back, I am generally happy with the selections fate made for me. There is one notable exception and that occurred in twelfth grade.

In my modest collection of mathematics books there is one that I have had for over fifty years. I've not read it in all that time. Perhaps once in a decade I pull it out and look inside the cover. There is the stamp that says "Return to Peabody High School," the high school in Pittsburgh that I attended. The textbook is *Elements of Calculus and Geometry* by George B. Thomas, Jr. This is perhaps my most prized book. It doesn't hold this status because of its content. It's the nature of its acquisition that sets this book apart.

German language and mathematics are intertwined in my life. If my academic path can be described as a success, these subjects formed the cornerstone. My cousin Grace lived with us in my maternal grandmother's house while she attended Chatham College in Pittsburgh. It was the mid-fifties and the institution had just adopted the new name. It had been called Pennsylvania College for Women. I was in the early years of grade school during the time that Grace spent with us and what she left with me when she graduated was a thirst for the German language. She was taking German at Chatham and she taught me some. It is also true that my father was stationed with the occupation troops in Germany right after World War II. He did learn some German, but I don't know if he ever spoke it in front of me when I was young. By the time I started grade school my parents had separated.

When it was time for me to go to high school, I petitioned the school board to allow me to attend a high school out of my district, because that high school offered German. It was also considered a better high school academically than the one I would have attended by default. I entered Peabody High School and excelled in German. It was as if it were a language I had once spoken and forgotten. As my relationship with German unfolded, it influenced in a positive way everything else I was doing academically.

However, I was not where I wanted to be in mathematics. I had opted not to accept the school board's assessment that my score on the placement test in algebra was good enough to skip Algebra I and begin geometry in the ninth grade. Our eighth-grade math teacher had missed almost the entire school year due to an automobile accident. All of the students selected for algebra that year had to teach themselves on their own time, while taking the regular arithmetic class with the other eighth-grade students. Two of us passed the placement test at the end of the school year, but despite this I balked at moving on to geometry. I felt I needed to go over the material again. My plan was to take Algebra II in the summer after tenth grade, so that I would be eligible to take calculus in twelfth grade.

My plan ended up being threatened by opportunity. I was pulled out of my tenth-grade chemistry class by the principal. In the hallway Mr. Mong told me that he was nominating me for the Yale Summer High School. It was the first year of the program and he had learned about it while attending a conference. I didn't even know where Yale was, although I was pretty sure it was in New England. My response was that I would like to do it, but I had other plans for the summer. I was going to take Algebra II. Mr. Mong smiled at me, a smile that clearly indicated that I did not understand. It was a smile that indicated that he was not going to let me turn down a summer in New Haven, if I was selected.

And I was selected to be a student in the first installment of the Yale Summer High School. With about one hundred other boys, I was housed on the secluded grounds of the Yale Divinity School. The participants were drawn from many different communities east of the Mississippi. I suppose it was a Yale experiment, taking kids with backgrounds that did not fit the prep school mold, boys from Appalachia, the inner cities, the rural south, New Haven's Italian community, ... It was a grand opportunity for all of us. We were immediately introduced to mathematics that we were unlikely to see in our high school curriculum and literature that we were unlikely to read. But I still worried about my plan, Algebra II.

At some point during the first week of the program I revealed my dilemma to my math instructor, George Cohan. He simply shrugged his shoulders and said he could teach me Algebra II on the side. He met with me two or three times a week for about five weeks and guided me through the material. Mr. Cohan wrote a letter to the Pittsburgh Board of Education describing the course that he had cobbled together for me. It was based on a one-year course that he taught at his high school in Westport, Connecticut.

After my return to Pittsburgh and shortly before classes began I received a letter from the Board of Education. They recognized the letter from George Cohan and advised me that I would need to take a placement test. It was cut and dry. I took the test and the result gave me credit for an "A" in Algebra II. Unlike the result of the placement test for Algebra I, I believed in this result.

While waiting for the School Board to pass its approval to my high school, I was parked in a section of Algebra II taught by Mr. Levy. I don't remember much about it, but I do remember Mr. Levy telling the class that this would be our first dose of real mathematics. Fortunately, Mr. Cohan had compressed my dose into less than six weeks. When my slip for transfer to trigonometry arrived, I handed it to Mr. Levy along with the Algebra II textbook, foreshadowing another exchange that would occur between us almost two years later.

German language held me together. It gave me an identity in the school. I was the German scholar. At times, I thought that I might chuck mathematics once I got to college and major in German. As you know, that did not happen. I took my last formal course in German, a literature course, my sophomore year in college, and that summer, in 1968, I went to Germany and worked at the Commerzbank in Düsseldorf in the Letter of Credits section. I thought it would be my capstone experience, but I wasn't through with Germany, not by a longshot.

I made it into the calculus class my senior year. Mr. Levy was the calculus teacher. We had been on a collision course. Mr. Levy was certainly far from being an optimal choice for me. A sequence of what I perceived as slights began. The tipping point occurred when I got a test back at the end of the grading period and saw a problem marked as incorrect that I knew was correct. It gave me a B on the test and that meant a B for the grading period. These were the grades that were going to be sent to colleges. I was applying to Yale and I needed to get into such a place, because I felt that it was only the wealthy institutions that were likely to give me the financial support that I needed to attend college. I had the not completely unfounded fear that failure to receive a substantial scholarship might get me drafted into the army. This was 1965 and boys from my neighborhood, Little Italy on the East End of Pittsburgh, were already getting shot at in Vietnam. I protested the mark. The answer was plus-or-minus some quantity. I presented my argument and Mr. Levy claimed that he did not recognize my plus-or-minus sign. I had not pulled the vertical piece of the sign down far enough to satisfy him. Instead of

I had written

I was furious. My classmate Arthur Denberg joined the argument. Mr. Levy was unmoved.

I acted up and I acted out. Every day for two weeks or more, before calculus class began, I would go to the chalkboard and write

Mr. Levy never gave an indication that he recognized my symbol of anger. There were other slights, but I held on. That spring I was admitted to Yale despite the "Levy B" in calculus. I don't want this to be a testament to standardized testing, but I am grateful for the scores on the SAT subject tests in mathematics and German. The letters of recommendation were written by my German teacher, Mrs. Thomas, and my eleventh-grade mathematics teacher, Ms. Walker. Ms. Walker was my trigonometry teacher, the one that Mr. Cohan delivered me to with his letter from Connecticut to the Pittsburgh Board of Education; the letter that allowed me to walk around Mr. Levy the first time we encountered each other.

At the end of the year textbooks are returned to your teacher. One by one we walked to the front of the room to hand in our calculus books to Mr. Levy. When it was my turn, I handed him my book and started to return to my seat. He looked up and said that it was not the book that he had issued me. The number in the book did not match the number on the card. I told him it was the book he had given me. Mr. Levy would not accept the book. In front of the class he said, "How do I know that you didn't lose your book and steal this one from Taylor Allderdice." Taylor Allderdice is a high school on the same side of the city. He told me that I would have to pay for the lost book. That meant I had to go home and ask my mother for five dollars, not a lot of money now and not really a lot of money then. But we had very little money. My mother earned a modest wage as an operating room technician. There was no use approaching my step-father. He was not supportive. He might have argued Levy's case.

A few days after my book was refused I went to the book repository and paid for it. The lady that ran the book room gave me a receipt and I again found myself in front of the class with Mr. Levy. I showed him the receipt to close the matter out. He looked up at me and asked for the book. I looked down at him and said, "I lost my book. You told me that." Then I returned to my seat. The logic was irrefutable. There were no questions about the correct formation of a symbol. I was becoming a mathematician.

* * *

I had four wonderful years at Yale, but I somehow lost my fire for mathematics. As the result of an event that can only be described as a miracle, I avoided the draft when I graduated in 1970. I had fully expected to be drafted and had not done any job interviews. On a lark, I joined two class-mates on a trip to Alaska. (It's amazing that not one of us has chosen to write a novel based on that summer.) We drove a camper from Tacoma to Fairbanks and then south to Anchorage, where we found work cleaning salmon. During the last six weeks, my home was in an abandoned pickup truck across the road from the Anchorage airport. I would sometimes sit in the airport terminal in the evening to read. Young soldiers, my age, would change planes there on the way to Vietnam. By mid-September, the nighttime temperatures dropped into the teens and my sleeping bag was inadequate.

Leaving my pickup truck, I flew south to the Lower 48 and landed a job as an economic analyst at Gulf Oil in Pittsburgh, where I labored for three years, paid off my mother's mortgage, and put aside quite a bit of money. Then, rather suddenly, I found myself in Germany again. I was drawn back. Eventually, I stumbled onto a job teaching algebra at one of the high schools in Düsseldorf and gradually realized that I should try to return to mathematics. The fire was reestablished. I left Germany after three years and began graduate school at the University of Pittsburgh a couple of months before my twenty-eighth birthday.

Before I started writing this chapter, I opened my treasured high school calculus book and found three pieces of paper that I did not know were there. One of them was a sheet of paper with some calculations I made long ago. There was also the receipt for the lost book. And there was a slip of paper from the International Correspondence Bureau in Munich, which I must have received when I was in tenth grade. On it was the name and address of the German pen pal that the organization had selected for me, Sigrid. As it turns out, my path to mathematics did pass through Germany. It seems appropriate that these slips of paper have resided together for more than fifty years in my old calculus textbook.

Chapter 8

Making the Optimal Choice

Very little of what appears in this book is closely related to what people usually call problem solving. I think the distinction is between delving into the structural nature of mathematics as a pursuit for its own sake as opposed to using mathematics to answer the day-to-day questions that human beings face. In the previous chapter I mentioned George Cohan, who taught me a year's worth of Algebra II in "ten easy lessons" on the Yale Divinity School campus during the summer of 1964. It is probably worthwhile to devote a sliver of space in this book to an application of mathematics that he taught us in the mathematics class that he conducted that summer. I feel that this application is particularly cute, and every young person with even a little bit of interest in mathematics should be familiar with it before graduation from high school. For this reason, I deign to give it Nursery Rhyme status.

A question that comes up all the time is: How do we maximize an objective given constrained resources? If you are talking about finding a spouse given your limited physical attractiveness, wealth, and social graces, mathematics is probably not going to help you. That is why I was not married until age forty-two. However, if you are a manufacturer trying to decide on how to maximize profit given constraints such as ability to pay labor costs and buy materials, mathematics can sometimes come to the rescue.

Now that I have made a close to serious pronouncement, I'm going to trot out an example that uses silly language. The reason for this is that I know nothing about manufacturing, which is the area from which the example is drawn. My use of silly language is a kind of confession right up front. Here we go.

Shirley has a factory that produces two products. One is called a widget. The other is called a gadget. She has constraints on how much she can spend on materials, delivery, and labor. She has translated these constraints into three inequalities. We let x represent the number of widgets she produces and y the number of gadgets. We'll write the right-hand sides of the constraint inequalities in multiples of 1000. For example, in the constraint immediately below, the amount of money that Shirley has to spend on materials is 160 thousand dollars:

$$C_1 : 5x + 8y \leq 160 \text{ (Material)},$$
$$C_2 : 7x + 4y \leq 116 \text{ (Delivery)},$$
$$C_3 : 8x + 3y \leq 120 \text{ (Labor)},$$
$$x \geq 0,$$
$$y \geq 0.$$

Shirley wants to maximize her profit and she knows exactly how her profit is computed. She earns \$7.00 profit for each widget and \$8.00 profit for each gadget. The profit P, also in thousands of dollars, is given by

$$P = 7x + 8y.$$

How many widgets and how many gadgets should she produce? Let's look at a picture of what the first constraint implies. The only *feasible* (doable given the constraints) solutions for x and y have to lie within the triangle and its boundary.

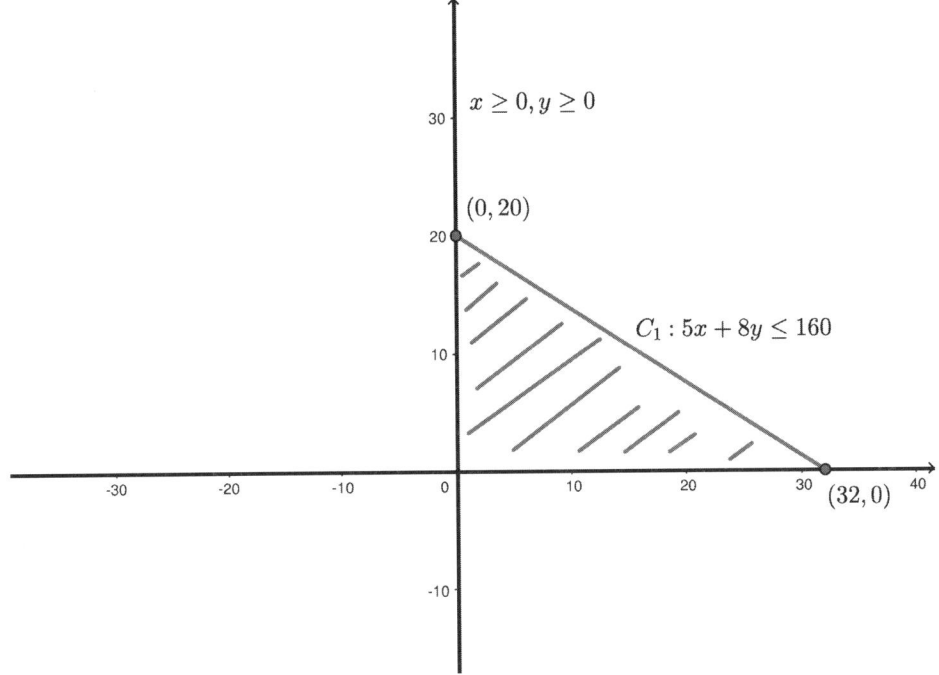

Of course, the points that are feasible, given only the constraint C_1, will be reduced when we add the additional constraints. When we do, we get a much smaller feasible region.

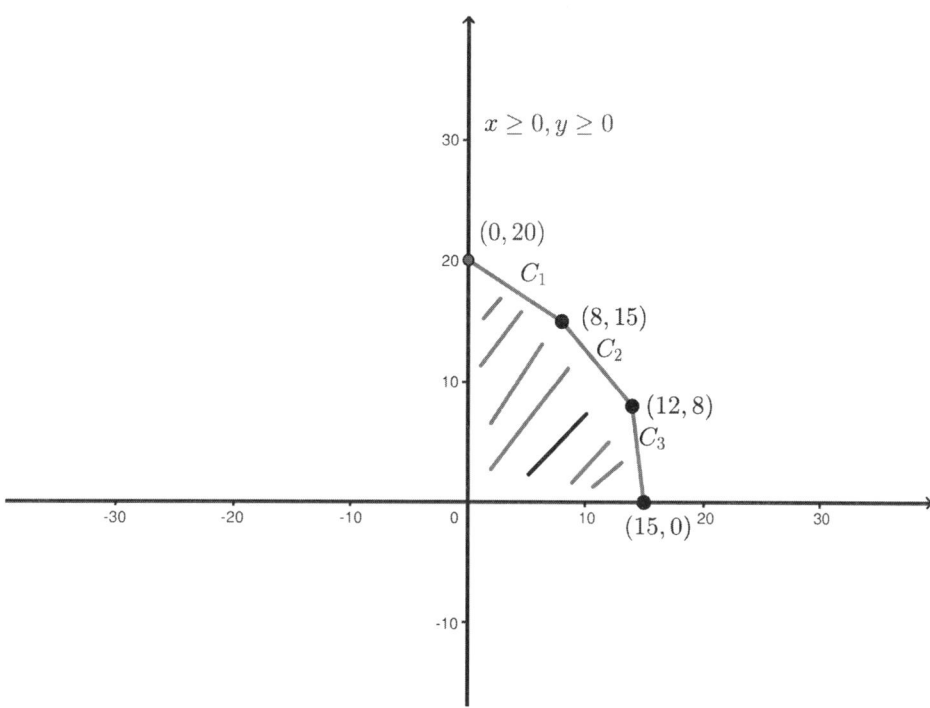

The picture above takes all three constraints into account. So how much money can Shirley make?

$$C_1 : 5x + 8y \leq 160 \text{ (Material)},$$
$$C_2 : 7x + 4y \leq 116 \text{ (Delivery)},$$
$$C_3 : 8x + 3y \leq 120 \text{ (Labor)},$$
$$x \geq 0,$$
$$y \geq 0.$$

She wants to maximize profit $P = 7x + 8y$. Gadgets, y, give her a higher profit rate than widgets, x, and the delivery and labor costs are lower. Should she bother with widgets at all?

First of all, we should check to see if Shirley can at least produce a profit high enough to stay in business. She mentioned to me that she will not continue in the business if she makes less than \$90,000 profit. We take a look at our picture to see if there are any points on the line

$$90 = 7x + 8y$$

that lie in the feasible region

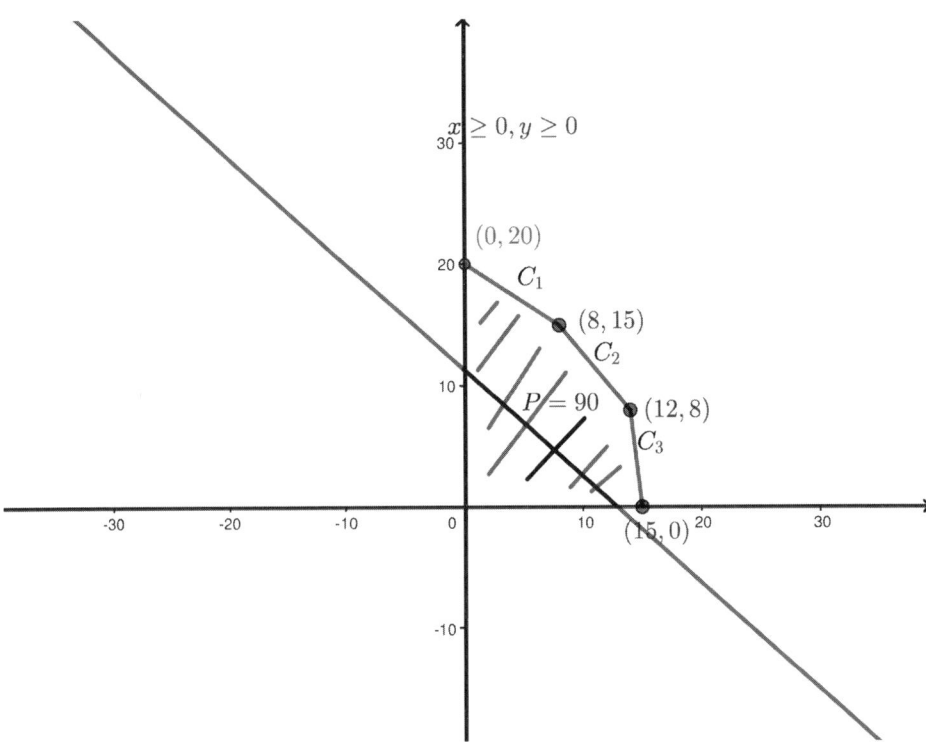

We see that there are many choices that will give her the minimum level of profit that she needs to stay in business. We also see that she can do a lot better. Each profit line will be parallel to the one that we drew for 90. To get a production mix that gives a greater profit we simply shift our profit line to the right keeping it parallel to the one we drew for the minimum profit she requires. When do we stop shifting? We stop shifting when any further shifting would cause us to lose touch with manufacturing reality. That is, when the line corresponding to the profit P no longer touches the shaded region. This means that we should stop when the line touches only a corner or perhaps contains an edge corresponding to one of our constraints on the right side of the feasible region. The possibility of it containing a whole edge would mean that the slope of the profit line is equal to the slope of one of the lines that corresponds to a constraint. In this case our profit lines are not parallel to any of the edges. This means we just have to test the corners on the rightmost boundary, and there are only four of them: $(0, 20)$, $(8, 15)$, $(12, 8)$, $(15, 0)$.

The profit is maximized by using the mix $(8, 15)$, which yields a profit of $176,000. That's 8,000 widgets and 15,000 gadgets. If she only manufactures gadgets, the best she could do is at $(0, 20)$, 20,000 gadgets, a profit of only $160,000. With hindsight, we see that all Shirley had to do in the first place was find out where the corners of the region were and test them with her profit function P.

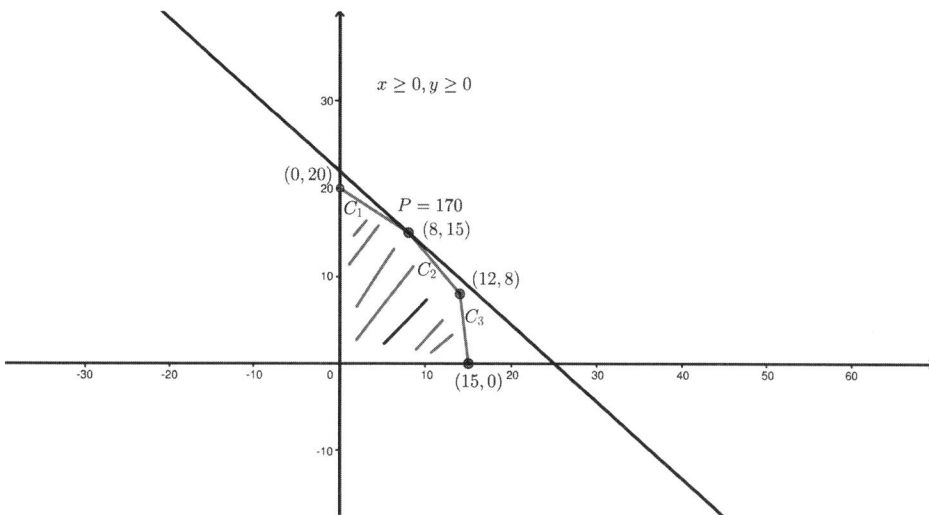

We also see that for this optimal solution Shirley will use all of the material and the delivery resources, but she will not have to use all of the labor resource. She has $120,000 labor dollars available and only uses $109,000. She can give someone a paid vacation.

We should ask about the case where the optimal point is not an integer and fractional solutions are not allowed. Well, then the problem can be difficult, especially if instead of a two-dimensional problem we have a problem in many dimensions. The best whole-number solution will, in general, not even be near the best fractional solution. The method that we have introduced in this chapter is called "Linear Programming". There are complex Linear Programming algorithms designed to tackle problems in higher dimensions. *Integer Programming* is what this class of problems is called when only integer solutions are admissible. Linear and Integer Programming are in the general area of mathematics called *Operations Research*.

Chapter 9

Impossibilities

A formula for the solution to the quadratic, essentially as we know it today, was likely first found by the Indian mathematician, Brahmagupta, in the seventh century. The pursuit of a formula for the solutions to the cubic equation,

$$ax^3 + bx^2 + cx + d = 0,$$

has a long history. A notable contributor on the path to its discovery was the poet-mathematician Omar Khayyám, who lived in the eleventh and twelfth centuries in Persia. Khayyám is best known among English speakers for the translations of his four-line poems, quatrains, by Edward FitzGerald, an example of which is

> The Moving Finger writes; and, having writ,
> Moves on: nor all thy Piety nor Wit,
> Shall lure it back to cancel half a Line,
> Nor all thy Tears wash out a Word of it.

It is useful to repeat what we said in Chapter 6. A formula in this context may only use the standard operations: addition, subtraction, multiplication, division, and radicals (root extraction, square roots, cube roots, etc.)

The quest for a cubic formula was successfully completed in the sixteenth century by the Italian mathematician, Gerolamo Cardano, who published a formula for the cubic together with the formula for the quartic, due to his student, Lodovico Ferrari.

Let's return to the quadratic for a moment. In Chapter 6 we derived the quadratic formula by completing the square. If completing the square works for the quadratic, could we perhaps simply just complete the cube for the cubic (third degree)? Attempting to complete the cube turned out to be a quixotic undertaking. The formula for finding a solution to the cubic is very painful. It's a dark and gloomy novella, not a nursery rhyme, and the quartic-formula story is worse. For more than two centuries after the cubic and quartic formulas were found, brave souls continued to pursue a formula for the quintic.

By the late 1700s, with no resolution in sight, Paolo Ruffini had the temerity to speculate that it might not be possible to produce a formula for solving the 5th degree polynomial equation using the standard operations. The assertion of the non-existence of a general formula is actually a stunning statement, and it met with considerable resistance, even outrage. In the early years of the 19th century two men were born, one in France and one in Norway, whose lives were short and tragic. Their work showed, in particular, the non-existence of a formula for finding solutions to polynomial equations of degree five or higher. The search for a quintic formula ended. There was no formula to be found. However, the work of Galois in France and Abel in Norway gave rise to whole branches of mathematics, which remain an important part of the formal training in the study of mathematics today. The parallels in the two men's lives are remarkable.

Évariste Galois was born in a suburb of Paris on October 25th, 1811. At age fourteen he opened a mathematics book written for experts and read it as if reading a novel. Mathematics had discovered him. Within a year Galois was reading the papers of the great mathematicians of the day. During this period his formal schoolwork declined from good to lackluster. Galois had other fish to fry. He was creating new mathematical structures and at the same time he became politically active, during a period when French politics were tumultuous. When he was eighteen his father committed suicide.

Galois was a headstrong young man. He clashed with school administrators. His efforts to bring the attention of the great mathematicians of the day to his work were largely unsuccessful, and when someone paid attention, he usually opted not to take the advice that he was given. Nonetheless, several of his papers found their way into publication after his death.

In his twentieth year Galois was arrested twice for belligerence in the political domain. The second time he spent six months in prison for illegally wearing the uniform of a disbanded National Guard artillery unit. He managed to continue his research while imprisoned.

A month after his release from prison Galois was killed in a duel, which was the apparent result of a difficult romantic relationship with a young woman. It has been suggested that he felt compelled to defend her honor. Anticipating that he would not survive the duel, Galois spent the night before the faceoff outlining the mathematical ideas that he feared he would not be able to carry to fruition.

Évariste Galois was shot in the abdomen and died the next day. His brother sat grieving at his bedside. Galois asked his brother Alfred to pull himself together. "Don't cry! I need all my courage to die at twenty."[1] He was buried in an unmarked grave.

On July 4, 1843, eleven years after Galois' death Joseph Liouville spoke before the French Academy:

"I hope to interest the Academy in announcing that among the papers of Évariste Galois I have found a solution, as precise as it is profound, of this beautiful problem: whether or not it is soluble by radicals. . ."[2]

---

[1] *Évariste Galois*, Laura Toti Rigatelli, Birkhäuser, p. 113.

[2] *Galois Theory*, Ian Stewart, Chapman and Hall, p. xv.

What follows is an excerpt from Duel at Dawn.[3]

The source of Galois' troubles, Dupuy believed, was that he lived as a stranger among his fellow men, even among his fellow patriots, as the Republicans of the time referred to themselves. "His true Fatherland and, the most beautiful and the largest of all, is necessarily the one where so many noble intelligences, dispersed in all corners of the world, fraternize among the rigorous conceptions and depths of mathematics." Galois' true home, in other words, was a universe separate from our own unpredictable world. It was the pure and perfect world of mathematics, as rational as it is beautiful, to which only a select few ever gain access. Condemned to live his life in the crass world of men rather than in the Platonic heaven of pure mathematics, Galois was doomed.

It is not rare to hear mathematicians lament the brevity of Galois's life, for what would such a genius not give us, if death had not taken him at twenty! But no. Galois, it seems, had fulfilled his destiny. If Galois had been admitted to the École Polytechnique, as he had hoped, he would have perished on the barricades of July 1830. And had he possessed sense enough to have avoided the duel two years later, he would certainly have died in the disturbances of 1832. Galois, according to Dupuy, was a stranger in our world and was marked for an early death.

Niels Henrik Abel was born in Nedstrand, Norway on August 5th, 1802. His father, Soren Abel, was a theologian and politician, who packed his son off at age thirteen to the Cathedral School in Oslo after determining that Niels' older brother was not yet emotionally prepared to leave home. Before Abel was sixteen years old his talent for mathematics had been discovered by one of his instructors. From that point on Abel was caught in the web. As he grew into the subject his father became embroiled in both political and theological disagreements with men of influence. Abel's father abandoned his membership in the Norwegian legislature and returned to his home, where he spent two years drinking himself to death. He died when his gifted son was eighteen years old. As you recall, this was Galois' age when his father committed suicide. By the time Abel was nineteen he was the most knowledgeable mathematician in Norway, despite the plunge into poverty that his father's death precipitated. When he matriculated at the university in Christiania (Oslo), the institution had little to offer him except time to read the research papers being produced by other mathematicians throughout Europe. It was then that he began his work on solving the quintic, which eventually led to his proving, four years later, the impossibility of producing a general formula for solving the quintic using the standard operations.

---

[3]*Duel at Dawn: Heroes, Martyrs, and the Rise of Modern Mathematics*, Amir Alexander, Harvard University Press, 2010, p. 97.

Abel's first attempt to go on a long tour of Europe beyond Scandinavia to meet the great mathematicians of the day was not supported. He was, however, given a stipend to spend two years learning French and German. In 1825, after directly petitioning the King of Norway-Sweden for permission to travel, Abel began a two-year jaunt through Europe, leaving wonderful mathematical results in his wake. In 1826, near the end of his travels, Abel reached Paris, where he lived for several months. He did not meet Galois, who was only fifteen years old at the time. Abel also did not take the time to visit the towering figure of his day, Carl Friedrich Gauss, although that had been an expectation when he was granted his travel stipend. He did contract tuberculosis. Abel was forced to return to Norway in 1827 having run out of funds and found no prospects for a mathematics position once he arrived home.

Abel died in 1928. He was twenty-six years old. He left behind a fiancée. Two days after his death a letter arrived from Berlin. It was from August Crelle, in whose *Journal of Pure and Applied Mathematics*, Abel had published quite a few research results. Crelle had secured him a position at the University of Berlin.

Felix Klein:

> But I would not like to part from this ideal type of researcher, such as has seldom appeared in the history of mathematics, without evoking a figure from another sphere who, in spite of his totally different field, still seems related ...I will not sound absurd if I compare his kind of productivity and his personality with Mozart's. Thus one might erect a monument to this divinely inspired mathematician like the one to Mozart in Vienna: simple and unassuming he stands there listening, while graceful angels float about, playfully bringing him inspiration from another world.

> Instead, I must mention the very different type of memorial that was in fact erected to Abel in Christiania and which must greatly disappoint anyone familiar with his nature. On a towering, steep block of granite a youthful athlete of the Byronic type steps over two greyish sacrificial victims, his direction toward the heavens. If need be, one might take the hero to be a symbol of the human spirit, but one ponders the deeper significance of the two monsters in vain. Are they the conquered quintic equations or elliptic functions? Or the sorrows and cares of his everyday life? The pedestal of the monument bears, in immense letters, the inscription ABEL.[4]

In 2003 the first Abel Prize was presented. The Abel Prize should be viewed as the equivalent of the Nobel Prize. Attempts to establish it had begun more than a century earlier. Nobel had not seen fit to establish a prize for mathematics.

---

[4]*Development of Mathematics in the 19th Century*, Math Science Press, 1979, p. 97.

So get this straight: There cannot be a general formula using standard operations to find the roots of polynomials of degree five or for any degree greater than five.

* * *

That was pretty heady stuff. What else is impossible? Well, impossibilities abound in the universe, as it turns out. Suppose I tell you that I have found a way to express the square root of two as the ratio of two integers. I hope to persuade you that anyone making such a claim is mistaken. In my case, since I know that it is not true, it would be more properly called lying.

We will assume that the ratio of two integers, a and b, where the two integers have no factors in common, equals the square root of two.

Then

$$\frac{a}{b} = \sqrt{2},$$

$$\frac{a^2}{b^2} = 2,$$

so

$$a^2 = 2b^2.$$

It's poetry time.

The product on the left side of the final equality has a factor of two, since the right side does. All prime factors that appear on the left side of the equation must appear an even number of times. (They appear twice as often in a^2 as they occur in a.) But since b and a have no factors in common, the right side of the equation, $2b^2$, can only have one factor of 2. That's an odd number of factors on the right. Our assumption has led to a contradiction. We must have assumed something that cannot happen, an impossibility.

Numbers that cannot be written as the ratio of two integers are said to be *irrational*, but bear no resemblance to human beings that are described using the same word.

Here's a claim that we are now in a position to tackle. There exist **irrational** numbers x and y such that

$$x^y$$

is **rational**.

Let x be $\sqrt{2}^{\sqrt{2}}$ and y be $\sqrt{2}$ and take a look at

$$x^y = \left(\sqrt{2}^{\sqrt{2}}\right)^{\sqrt{2}} = \sqrt{2}^2 = 2.$$

Remember that we have just shown that $\sqrt{2}$ is irrational. Now, there are two possibilities. Either x is irrational or it is not. If it is rational, it demonstrates all by itself that a rational number can be written as an irrational number raised to an irrational power. If x is irrational, then x raised to the square root of two is equal to two, a rational number.

Note that we have not shown which one of the cases is true, but one of them must be. Either one of the only two possibilities supports the claim. This argument reminds me very much of Euclid's proof of the infinitude of the prime numbers that we presented in the first chapter.

As it turns out, a rather deep theorem due to the 20th-century mathematicians Gelfond and Schneider, the Gelfond-Schneider Theorem, tells us that $\sqrt{2}^{\sqrt{2}}$ is not rational. So, in fact, $x = \sqrt{2}^{\sqrt{2}}$ and $y = \sqrt{2}$ demonstrate that there are irrational numbers that do the job. However, we did not need the heavy-duty Gelfond-Schneider hammer simply to prove the *existence* of two irrational numbers x and y such that x^y is rational.

$$* * *$$

Given that I claimed that there is a long list of impossibilities, it seems appropriate to offer one more example. This one concerns how distances can be arranged. If you are asked to place three points in the plane so that all of the distances are odd whole numbers, you can do it easily by simply drawing an equilateral triangle with all of its edges length 1. If that is not interesting enough, a triangle with sides of lengths 3, 5, and 7 will also work. The angle across from the length-7 side will be 120°. If more is demanded and you are asked to place four points in the plane with the distance between each pair of them being an odd integer, you have been asked the impossible.

Why can't we do it? I will try to answer this, but when I am done, you might accuse me of pulling a rabbit out of a hat, even though I am not a magician.

Despite what I just said, we proceed as if it is possible to place four points in the Cartesian plane so that all of the pairwise distances are odd integers. We will see that this leads to a logical contradiction.

We might as well assume that one of the points is at the origin. These are the four points:

$$O = (0,0), A = (a_1, a_2), B = (b_1, b_2), C = (c_1, c_2).$$

Denote the square of the distance from point x to point y by $d^2(x, y)$. Then:

$$d^2(O, A) = a_1^2 + a_2^2,$$
$$d^2(O, B) = b_1^2 + b_2^2,$$
$$d^2(O, C) = c_1^2 + c_2^2.$$

If we are going to pull a rabbit out of a hat, we will need a hat. Now we do something that is completely unmotivated to create a "hat." We look at the following array:

$$\begin{bmatrix} a_1^2 + a_2^2 & a_1 b_1 + a_2 b_2 & a_1 c_1 + a_2 c_2 \\ a_1 b_1 + a_2 b_2 & b_1^2 + b_2^2 & b_1 c_1 + b_2 c_2 \\ a_1 c_1 + a_2 c_2 & b_1 c_1 + b_2 c_2 & c_1^2 + c_2^2 \end{bmatrix}.$$

There is something interesting about this array. I can find a formula for writing the third row as a multiple of the first row plus a multiple of the second. Actually, I found a formula before I started writing. (This is like one of those television cooking shows.) If we multiply the top row and the middle row by

$$\frac{b_1 c_2 - b_2 c_1}{a_2 b_1 - a_1 b_2} \quad \text{and} \quad \frac{a_2 c_1 - a_1 c_2}{a_2 b_1 - a_1 b_c},$$

respectively, and then add them, we get the bottom row. We should mention that the denominator in the two expressions immediately above cannot be zero, since that would imply that

$$\frac{a_2}{a_1} = \frac{b_2}{b_1}.$$

That in turn would mean that the points A, B, and O are co-linear. But three points cannot be on the same line if all three of their pairwise distances are odd.

For those unfamiliar with modular arithmetic, a brief introduction is provided in Appendix A.

Now we ready ourselves for some pulling of a fairly heavy rabbit. We note that if we divide the square of any odd integer by eight, the remainder is one. (We say that the square of an odd integer is 1 modulo 8.) This is seen by squaring the odd integer $2n + 1$:

$$(2n + 1)^2 = 4n^2 + 4n + 1 = 4n(n + 1) + 1.$$

Since either n or $n + 1$ is even, the first term in the rightmost expression must be divisible by eight.

If we look at our array, we see that the diagonal is simply the square of the distances of points A, B, and C to O. All of the distances are assumed to be odd. This tells us that all three entries on the diagonal of our array can be written in the form $8s + 1$, for some integer s.

The following three equalities are useful:

$$d^2(A, B) = (a_1 - b_1)^2 + (a_2 - b_2)^2$$
$$= a_1^2 + a_2^2 + b_1^2 + b_2^2 - 2(a_1 b_1 + a_2 b_2)$$
$$= d^2(O, A) + d^2(O, B) - 2(a_1 b_1 + a_2 b_2),$$
$$d^2(A, C) = d^2(O, A) + d^2(O, C) - 2(a_1 c_1 + a_2 c_2),$$
$$d^2(B, C) = d^2(O, B) + d^2(O, C) - 2(b_1 c_1 + b_2 c_2).$$

Using the second equality as an example, we can claim the following is true for some integers f, g, and h, because all the squared distances on the right side of the top equation are equal to 1 mod 8:

$$2(a_1 c_1 + a_2 c_2) = d^2(O, A) + d^2(O, C) - d^2(A, C)$$
$$= (8g + 1) + (8h + 1) - (8f + 1) = 8(g + h - f) + 1.$$

That tells us that if we double the off-diagonal elements of the array, the result is also an integer that is 1 modulo 8. Since the diagonal elements are of the form $8s + 1$, for some integer s, doubling them gives us integers of the form $8t + 2$, for $t = 2s$.

In summary, doubling all of the elements in the array would give us an array with integers that are 2 modulo 8 on the diagonal and 1 modulo 8 off the diagonal. We depict it this way with m, n, k, j, a, and b representing integers:

$$\begin{bmatrix} 8m + 2 & 8k + 1 & 8a + 1 \\ 8k + 1 & 8n + 2 & 8b + 1 \\ 8a + 1 & 8b + 1 & 8j + 2 \end{bmatrix}.$$

Modifying the array in this way should not change what we established in the hat-construction phase of this discussion. The bottom row must still be the sum of multiples of the top two. Here comes the rabbit: **It is not!** It does not matter how the integers m, n, k, j, a, and b are chosen.

Being heroic, I found numbers x and y that satisfied the system below, which uses the first two entries in each row:

$$x(8m + 2) + y(8k + 1) = 8a + 1,$$
$$x(8k + 1) + y(8n + 2) = 8b + 1.$$

The solution is

$$x = \frac{1 + 16a - 8b - 8k - 64bk + 8n + 64an}{3 - 16k - 64k^2 + 16m + 16n + 64mn},$$
$$y = \frac{1 - 8a + 16b - 8k - 64ak + 8m + 64bn}{3 - 16k - 64k^2 + 16m + 16n + 64mn}.$$

(Yes, you can carry out these computations too. An old man did.)

I've used only the first two elements in each row. What happens if I try to find the solution using the second and third elements of each row? I might expect to get something equivalent. That would be the solution to the following system:

$$x(8k + 1) + y(8n + 2) = 8b + 1,$$
$$x(8a + 1) + y(8b + 1) = 8j + 2.$$

The corresponding solution is

$$x = \frac{3 - 16b - 64b^2 + 16j + 16n + 64jn}{1 + 16a - 8b - 8k - 64bk + 8n + 64an},$$
$$y = \frac{1 - 8a - 8b - 64ab + 8j + 16k + 64jk}{-1 - 16a + 8b + 8k + 64bk - 8n - 64an}.$$

We could imagine that for some choice of the variables these two solutions are the same. Let's question if this is true in the case of the two expressions for y. We can set them equal to each other and (without panicking) note that when we cross-multiply, the terms on both sides of the resulting equation are divisible by 8 except for the "3" term on one side and the "-1" term on the other. (Remember that m, n, k, j, a, and b are all integers.) Writing it in this high-level way (no need to look at the individual trees, look at the forest) we have

$$-1 + 8M = 3 + 8N$$

or

$$8M - 8N = 4$$

where M and N are integers. Well, this is impossible, because it would imply that 4 is divisible by 8. This shows that the third row cannot be written as the sum of multiples of the first two. This is at odds with the fact that we have also shown that this can be done.

So, we find ourselves in the Land of Contradiction, which is due entirely to our assumption that all six of the distances are odd integers. We are forced to conclude that it is impossible to find four points in the plane with all pairwise distances odd integers. The proof using the machinery of linear algebra is a lot slicker, but no less mysterious. In fact, the proof that I have just presented is a repackaging of that proof using elementary methods. If you are interested *and* know some linear algebra, take a look at the slick proof in Appendix C.

Chapter 10

Magnitudes of Infinity

I took a course in set theory in college. The professor was Abraham Robinson. Throughout my undergraduate years I was remarkably ignorant when it came to the reputation of my professors. Robinson is considered the father of non-standard analysis, a major 20th-century figure in the foundations of mathematics. What I did know was that he would call on me without warning, always smiling, "Melvin?" I had very little in my head. That was many decades ago, but I do not have the feeling that I ever gave a satisfactory response. He was good to me. I passed the course. I am now seventeen years older than he was when he stood in front of us at Yale. I wonder how it was that he even knew my name. It was a large class.

Five years later, I found myself teaching algebra at a public school in Düsseldorf, die Kaufmännische Schule IV. I picked up Fraenkel's *Abstract Set Theory*, and read it cover to cover. I read it the way I should have read it when I was carrying it back and forth to Robinson's classroom. It was 1974 and I was in Germany, where Robinson was born. It was also the year he died at age fifty-five.

My goal is to talk about magnitudes of infinity in a way that is comprehensible. I'm not greatly concerned that it be presented with the starch that one would find in the writing of a luminary such as Fraenkel. After all, we are doing nursery rhymes.

I am shooting for understanding. The first thing we need to do is think about how we count. (You might remember that I said the same thing on the way to proving the Pythagorean Theorem.) Counting will lead to an understanding that infinitudes are not all equal. As a result, we will be convinced of the existence of transcendental numbers, a special breed of the irrational numbers, which we discussed earlier.

We say that a number is *algebraic* if it is the root of a polynomial with integer coefficients. That is, a number is algebraic just in case it satisfies an equation of the form

$$a_n x^n + a_{n-1} x^{n-1} + \cdots + a_1 x + a_0 = 0, \text{ where each } a_j \text{ is an integer, } 0 \le j \le n,$$

for some positive integer n. A number that is not algebraic is said to be *transcendental*. So how do we find a transcendental number? Sometimes they introduce themselves to us. One of them introduced itself to various civilizations, our friend π. Another transcendental number is e, the base for what is called the natural logarithm. These numbers were shown to be transcendental by two 19th-century mathematicians, Hermite (e) in 1873 and Lindemann (π) in 1882. Before them Liouville (the man who helped pull Galois from obscurity) showed in 1844 that the number

$$\sum_{k=1}^{\infty} 10^{-k!}$$

is transcendental. However, e and π were the first "naturally occurring" numbers shown to be transcendental.

Essentially, this whole chapter rests on the work of Georg Cantor (1845-1918), who devised and developed the theory of sets. He gave us an amazing proof of the existence of transcendental numbers without using the heavy and difficult machinery of analysis. Cantor simply counted in an ingenious way. While counting, he opened a door that revealed an infinity of infinities. We shall see that some infinities are larger than others.

We have an abstract system of numbers. When the numbers get so big that we run out of names, we still have their symbolic representation. We use this system to measure how many objects we have before us. Using the system, we can say if we see more of some objects today than we saw yesterday. We can compare the number of apples with the number of oranges.

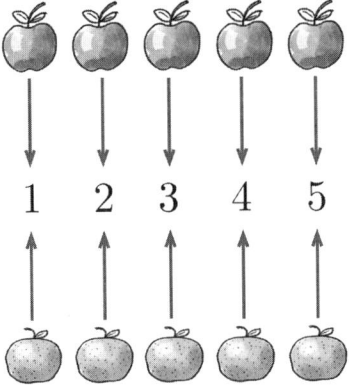

Our investigation will focus on whether or not two sets of objects have the same *cardinality* (size). Without our elaborate numbering system, we would be forced to compare the two sets directly and establish whether or not we could pair one element of the first set with precisely one element of the second and vice versa. We would have to decide if there is a one-to-one (1–1) correspondence between the two sets.

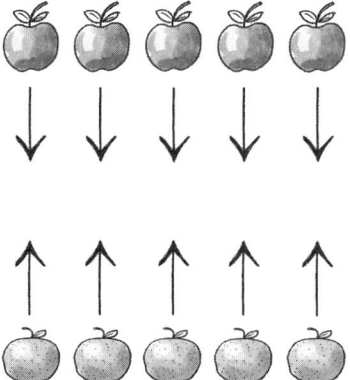

The numbering system as middleman goes away and we are left with this crude matching, but it is this crude matching that we will need in our attempt to make sense of magnitude when discussing infinite sets.

It is easy to see that there is a 1–1 correspondence between the positive integers and the positive even integers. We simply match each positive integer n with its double, $2n$. We learned earlier that there are infinitely many prime numbers. We can then form a 1–1 correspondence between the positive integers and the primes by matching the nth integer with the nth prime. We can then say that the set of all positive integers is the same cardinality as the set of all primes and the set of all positive even integers. We see an interesting distinction between the infinite set of positive integers and finite sets. There cannot be a 1–1 correspondence between a finite set and any of its *proper* subsets. We have to say proper subset, because every set is considered to be a subset of itself, but the whole set is not a proper subset.

We can easily write a 1–1 correspondence between the set of positive integers and the set of all integers. Here is one way to do it. For $n \geq 1$ match each positive even integer $2n$ with the integer n and each odd positive odd integer $(2n-1)$ with $1-n$. Here are a few pairings in this correspondence:

$$
\begin{array}{ccc}
1 & \to & 0, \\
2 & \to & 1, \\
3 & \to & -1, \\
4 & \to & 2, \\
5 & \to & -2, \\
6 & \to & 3 \\
& \vdots &
\end{array}
$$

Now we compare the positive integers with the positive rational numbers. Recall that a number is rational if it can be written as the quotient of integers. Well, the rationals are a lively bunch, unlike the integers and primes. Between any two of them you can always find another one. We cannot use their relative sizes to line them up and fashion a 1–1 correspondence with the positive integers. However, we can follow the arrows in the following graphic and easily establish a correspondence that way.

1	2	3	4	5	6	7	8
$\frac{1}{1}$	$\frac{1}{2}$	$\frac{1}{3}$	$\frac{1}{4}$	$\frac{1}{5}$	$\frac{1}{6}$	$\frac{1}{7}$	$\frac{1}{8}$
$\frac{2}{1}$	$\frac{2}{2}$	$\frac{2}{3}$	$\frac{2}{4}$	$\frac{2}{5}$	$\frac{2}{6}$	$\frac{2}{7}$	$\frac{2}{8}$
$\frac{3}{1}$	$\frac{3}{2}$	$\frac{3}{3}$	$\frac{3}{4}$	$\frac{3}{5}$	$\frac{3}{6}$	$\frac{3}{7}$	$\frac{3}{8}$
$\frac{4}{1}$	$\frac{4}{2}$	$\frac{4}{3}$	$\frac{4}{4}$	$\frac{4}{5}$	$\frac{4}{6}$	$\frac{4}{7}$	$\frac{4}{8}$
$\frac{5}{1}$	$\frac{5}{2}$	$\frac{5}{3}$	$\frac{5}{4}$	$\frac{5}{5}$	$\frac{5}{6}$	$\frac{5}{7}$	$\frac{5}{8}$
$\frac{6}{1}$	$\frac{6}{2}$	$\frac{6}{3}$	$\frac{6}{4}$	$\frac{6}{5}$	$\frac{6}{6}$	$\frac{6}{7}$	$\frac{6}{8}$
$\frac{7}{1}$	$\frac{7}{2}$	$\frac{7}{3}$	$\frac{7}{4}$	$\frac{7}{5}$	$\frac{7}{6}$	$\frac{7}{7}$	$\frac{7}{8}$
$\frac{8}{1}$	$\frac{8}{2}$	$\frac{8}{3}$	$\frac{8}{4}$	$\frac{8}{5}$	$\frac{8}{6}$	$\frac{8}{7}$	$\frac{8}{8}$

You see that we do skip over any fraction that is not reduced to lowest terms. By doing this we assure that each rational appears only once on our list. Our arrows will visit each positive rational and create, at each step, a match with one and only one positive integer:

$$1 \to \frac{1}{1},$$

$$2 \to \frac{2}{1},$$

$$3 \to \frac{1}{2}$$

$$\vdots$$

We say that a set is *countably infinite* if a 1–1 correspondence exists between the set and the positive integers. We sometimes simply say that the set is countable if it is already clear that the set is infinite. We mention here that the above proof, showing that the set of rational numbers is countable, can easily be adapted to show that the union of finitely or even countably many countable sets is also countable. This is done by listing each of the countable sets as a vertical column (horizontal would also work) and, to use a precise term, "zig-zagging" as we did in the table above, eventually reaching every element in each set, and discarding any element that has already been assigned. Discarding will be necessary if the sets have elements in common.

So far, we have not produced an infinite set that is not countable. Next, we consider the algebraic numbers. The reader's attention is drawn to the fact that the sets that we have been discussing consist entirely of algebraic numbers. Every integer satisfies

$$x + a_0 = 0,$$

for some integer a_0. Every rational number satisifies

$$a_1 x + a_0 = 0,$$

for some integers a_1 and a_0.

The only number whose irrationality we have discussed explicitly, $\sqrt{2}$, is also algebraic, since it satisfies

$$x^2 - 2 = 0.$$

Have we finally found an infinite set that cannot be put in 1–1 correspondence with the integers? No, we have not. The fact that **a polynomial of degree n has at most n distinct roots** is the essential piece that allows us to show that the algebraic numbers can be put into 1–1 correspondence with the positive integers.

Here's a device that helps us get organized to construct the 1–1 correspondence between the positive integers and the algebraic numbers. We define the height, h, of our polynomial with integer coefficients

$$a_n x^n + a_{n-1} x^{n-1} + \cdots + a_1 x + a_0$$

to be

$$h = n + |a_n| + |a_{n-1}| + \cdots + |a_1| + |a_0|, \text{ with } |a_n| \geq 1.$$

Slowing down for a moment, we see that there are only finitely many polynomials that have a given height, h. That is because all of the summands in the summation that determines the height are positive and there are only finitely many choices of positive integers (partitions) that sum to a given integer. We have to account for the fact that we are taking the absolute value of each coefficient and not necessarily the coefficient itself, but, even allowing for that, there are only finitely many polynomials for any given height.

For a given height h, each of the finitely many polynomials that have that height has only finitely many roots, in particular only finitely many roots that are real numbers. So only finitely many real algebraic numbers correspond to each h. For each h, starting with the smallest possible one, 2, we order all of the roots according to any rule that we wish, for instance magnitude. For height 2, there is only one root, 0, since we have only two polynomials of height 2, namely $\pm x$.

So we begin our 1–1 correspondence with the positive integers by matching 1 with 0.

$$1 \to 0.$$

The equations for polynomials of height 3 are

$$x + 1 = 0,$$
$$-x - 1 = 0,$$
$$x - 1 = 0,$$
$$-x + 1 = 0,$$
$$2x = 0,$$
$$-2x = 0,$$
$$x^2 = 0,$$
$$-x^2 = 0.$$

We only pick up two new algebraic numbers, 1 and -1. The root, 0, was assigned when we considered height 2.

We extend our 1–1 correspondence to

$$1 \to 0,$$
$$2 \to -1,$$
$$3 \to 1.$$

For height 4, we list only the equations for polynomials with positive coefficients for the sake of brevity:

$$x^3 = 0,$$
$$x^2 + x = 0,$$
$$2x^2 = 0,$$
$$x^2 + 1 = 0,$$
$$x + 2 = 0,$$
$$2x + 1 = 0,$$
$$3x = 0.$$

We note that the equation $x^2 + 1 = 0$ has no real roots. As we work through the rest of the list discarding all of the real algebraic roots that we have already found and also examining the remaining height-4 polynomials, we harvest four new algebraic numbers:

$$\pm \frac{1}{2} \text{ and } \pm 2.$$

These four new values extend our correspondence to

$$1 \to 0,$$
$$2 \to -1,$$
$$3 \to 1,$$
$$4 \to -2,$$
$$5 \to -\frac{1}{2},$$
$$6 \to \frac{1}{2},$$
$$7 \to 2.$$

We continue on our merry way in this fashion. At the next height, 5, the square root of two will get listed, since it is a root of the height-5 polynomial $x^2 - 2$. So we see that the set of algebraic numbers is countable.

At this point one might conjecture that the set of real numbers is also countable, if I had not indicated at the beginning of this chapter that there is a punch line. I remind the reader of the reasoning that emerged in the discussion of the infinitude of primes. I said that if we can show that every finite set of primes is missing at least one prime, then the sequence of primes must be infinite. There is an analogous statement that we can apply in our current situation. If we can show that every countably infinite set is missing at least one real number, then the real numbers cannot be countable. And that is what we now show.

We can list the elements of a countable set S of numbers by using the decimal representation for each element. We choose the decimal representation that has infinitely many non-zero digits after the decimal point. Of course, $1/3$ has only one decimal representation, but some numbers have two.

$$\frac{1}{8} = 0.125 = 0.12499999\ldots.$$

The listing below effectively provides an attempted 1–1 correspondence between $\mathbf{R}$ (the set of real numbers) and the positive integers. Here is our list:

$$a_1.a_{11}a_{12}a_{13}a_{14}a_{15}a_{16}a_{17}\cdots,$$
$$a_2.a_{21}a_{22}a_{23}a_{24}a_{25}a_{26}a_{27}\cdots,$$
$$a_3.a_{31}a_{32}a_{33}a_{34}a_{35}a_{36}a_{37}\cdots,$$
$$a_4.a_{41}a_{42}a_{43}a_{44}a_{45}a_{46}a_{47}\cdots,$$
$$a_5.a_{51}a_{52}a_{53}a_{54}a_{55}a_{56}a_{57}\cdots.$$

The digit before the decimal point, which includes the minus sign, if the number is negative, will not play a role in the construction. We build a number that is not on this list by exploiting the values on the diagonal. The jth decimal place of our number is taken to be 2, if a_{jj} is not 2, otherwise the jth decimal place is taken to be 1. The resulting number is not on the

list, since the number that we have built has an infinite expansion with no
alternate terminating decimal expansion, and its expansion is different from
the infinite expansion of every number on the list:

$$a_1.\mathbf{a_{11}}a_{12}a_{13}a_{14}a_{15}a_{16}a_{17}\cdots,$$

$$a_2.a_{21}\mathbf{a_{22}}a_{23}a_{24}a_{25}a_{26}a_{27}\cdots,$$

$$a_3.a_{31}a_{32}\mathbf{a_{33}}a_{34}a_{35}a_{36}a_{37}\cdots,$$

$$a_4.a_{41}a_{42}a_{43}\mathbf{a_{44}}a_{45}a_{46}a_{47}\cdots,$$

$$a_5.a_{51}a_{52}a_{53}a_{54}\mathbf{a_{55}}a_{56}a_{57}\cdots$$

$$\vdots$$

Since the set of algebraic numbers is countable, there must be real numbers that are not algebraic. This argument shows the existence of transcendental numbers without producing one. The set of transcendental numbers
must be larger than the set of algebraic numbers. If the set of transcendental numbers (non-algebraic numbers) were countably infinite, it would follow
that the real numbers, the union of the algebraic numbers and the transcendental numbers, must be countable, and we have just shown the contrary.
It then makes sense to say that *most* real numbers are transcendental. Sets
that are neither finite nor countably infinite are said to be *uncountable*. The
real numbers are uncountable. The reasoning that allowed us to conclude
that the set of real numbers cannot be put in 1–1 correspondence with the
set of positive integers is called *Cantor diagonalization*, and generalizations
of this technique are found in many branches of mathematics.

We can say something rather outrageous. If it were possible to put in
place a mechanism that selected a real number in the interval $[0, 1]$ truly
randomly, the probability of that number being algebraic is zero! Here's one
way to generate some intuition for this statement. Imagine that you could
select a positive integer randomly from the set of positive integers. What's
the probability that you select the number 1? Well it has to be less than
$1/1000$, because that is the probability if there were only 999 other equally
probable selections. In fact, by the same reasoning, it must be less than $1/n$
for any choice of n distinct integers, no matter how large. It must be zero.
This was a selection of a finite set from a countably infinite one.

Now what about the probability of getting an algebraic number when
choosing a number randomly from the interval $[0, 1]$? We'll let ϵ be a very
small positive number. Since the algebraic numbers are countable, we can
put them in one-to-one correspondence with the positive integers. (We have
already shown this for the entire set of algebraic numbers, and it is certainly
true if we restrict our attention to the algebraic numbers in the interval
$[0, 1]$.) We can list the algebraic numbers sequentially, a_1, a_2, a_3, etc. Let
I_k be the interval of length $\frac{1}{2^k}\epsilon$ centered at a_k. If the interval I_k contains

points outside the interval $[0, 1]$, reduce it in size so that it fits inside $[0, 1]$ and still contains a_k. We conclude that

$$\sum_{k=1}^{\infty} \text{length}(I_k) \leq \sum_{k=1}^{\infty} \frac{1}{2^k}\epsilon = \epsilon.$$

Remember that ϵ is small. This means that we have missed a lot of the interval $[0, 1]$. Almost all of it. (If you, the reader, can visualize this, I fear for your sanity.) Were we able to physically grab a number in $[0, 1]$, the chance of getting an algebraic number is certainly no more than ϵ. But ϵ is as small as we want to make it. So our chances are less than any positive number. *Zero!* Notions such as this are handled carefully in a subject called *measure theory*.

On a philosophical note, one of the most pleasing results of Cantor's work is the proof that there is always a magnitude of infinity larger than the largest one you have found. In fact, what is true for finite sets is also true for infinite sets: The set of all subsets of a set has larger magnitude than the set itself. We will not prove that in this book, but we show in Appendix D that this is true for the set of all subsets of the positive integers.

Chapter 11

The Inevitable (Sperner's Lemma—The Brouwer Fixed-Point Theorem)

I do not remember when I first encountered Sperner's Lemma. I do know that it was when I was well beyond the nursery-rhyme phase of my mathematical development. I wish I had bumped into it much earlier. It's a real delight. This is how the story is told.

Choose an arbitrary triangle and label the vertices A, B, and C.

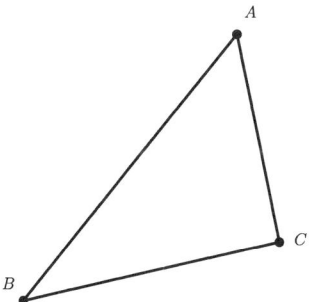

Next, we subdivide each edge as follows. On edge AB, we arbitrarily pick a finite subset of points and randomly assign each point either the letter A or the letter B. Neither the number of points we pick nor the letter (A or B) assigned to the chosen points matters. We take the analogous action on edges BC and AC. Below is a picture of our triangle after a subdivision of its edges.

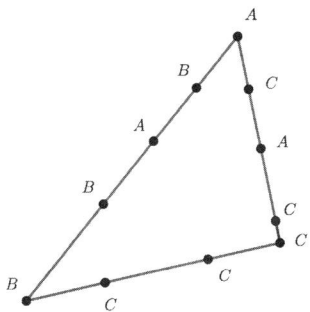

Now we subdivide the interior of the triangle with smaller triangles, subject to the requirement that any edge of a small triangle be either one of the edges produced by our subdivision in the triangle above or, if it is in the interior of the original triangle, it must be the edge of exactly two of the small triangles. This is called a *triangulation* of the triangle ABC. When this task is completed we arbitrarily label each vertex in the interior of our original triangle with any letter A, B, or C. Below is a possible final product.

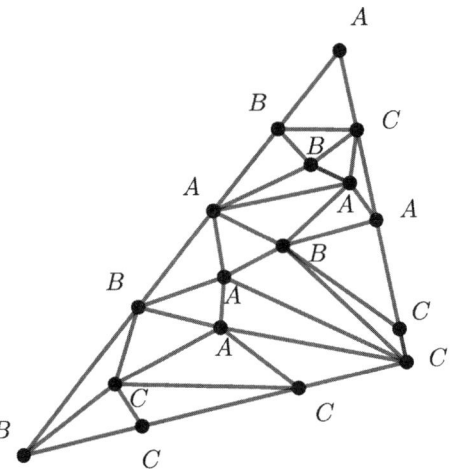

When we check to see how many of the triangles in our triangulation have all three possible vertex names, A, B, and C, we find that five of them do. We will call triangles that have all three letters labeling their vertices *complete* triangles. We mark the complete triangles below and explain the arrows in the next paragraph.

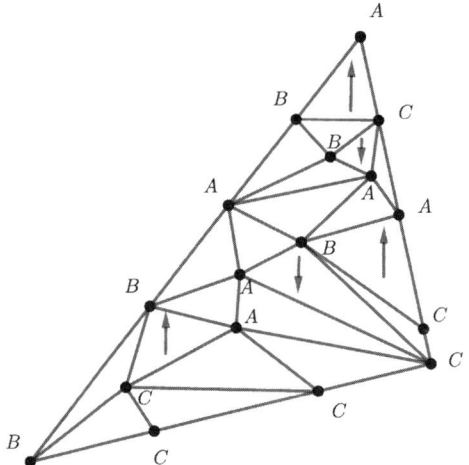

We notice that three of the interior complete triangles have the same orientation (up arrows) as the vertices A, B, C of the large outer triangle. If we follow A to B to C and back to A, we go counterclockwise. Two have the opposite orientation (down arrows), where following the vertices A to B to C and back to A we go clockwise. Surprisingly, Sperner's Lemma tells us that any triangulation will always produce at least one complete triangle. It tells us even more. We will always have one more complete triangle with the same orientation as the original triangle than we have complete triangles with the opposite orientation. Of course, we are going to try to show that this claim is true.

We will say that the *orientation* is $+1$ when moving along edge AB from A to B, along BC from B to C, and finally along CA from C to A is counterclockwise. If we imagine walking the perimeter counterclockwise, the interior of the triangle will always be on our left, as in the triangle below.

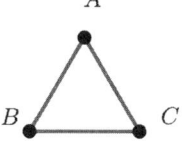

If the motion around the perimeter, A to B to C, is clockwise, the orientation is said to be -1.

It is useful to define the orientation of the edges AB, BC, and CA to be $+1$ if the triangle ABC has a counterclockwise orientation and -1 otherwise. That is to say that the triangle's edges inherit the orientation of the triangle in a straightforward way.

Now consider the triangle below. We have subdivided edge AB by adding a point and labeling the point A.

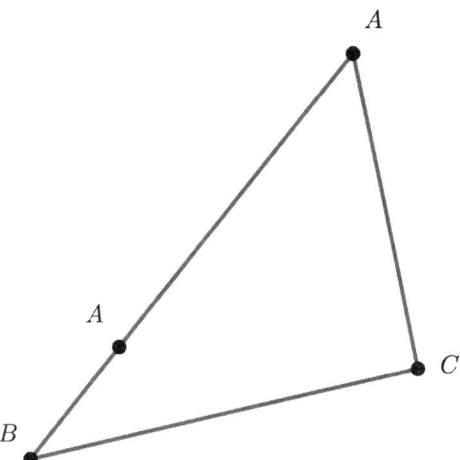

Our original AB edge now consists of two segments, AA and AB. We define the orientation of the segment AB to be the same as the orientation of the edge if the direction from A to B on the segment is the same as the direction from A to B of the vertices A and B. We see that the direction of the segment AB is the same as the direction of the directed line segment from vertex A to vertex B. So both edge and segment have orientation $+1$.

Similarly, if we subdivide our edge AB, by placing a B label between A and B, we are again left with only one segment AB with the $+1$ orientation.

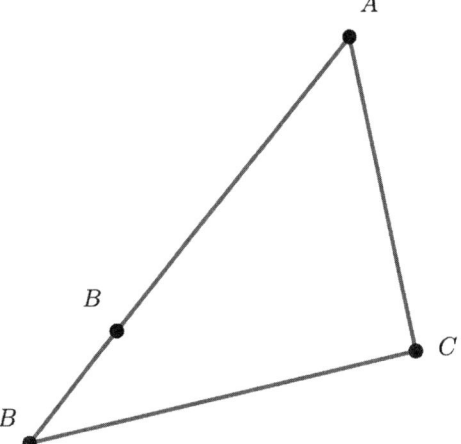

In general, we define the *Index* of the edge AB, which we have subdivided, to be the sum of the orientations of the segments that have both labels, A and B. For example, the index of the subdivided edge below is $+1$.

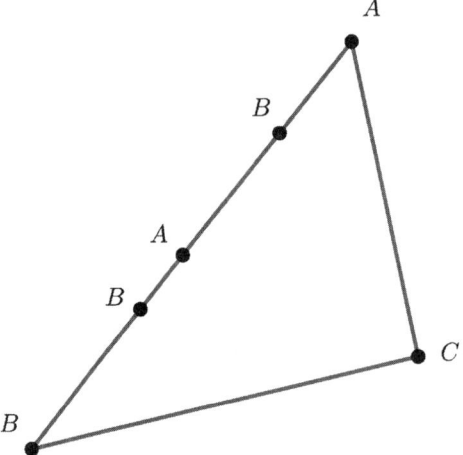

We see this by moving down the edge from right to left: we have an AB segment with orientation $(+1)$, BA segment (-1), and AB segment $(+1)$. The final segment, BB, does not contribute to the Index. The Index, the sum of the orientations, is $+1$.

It is now straightforward to see that the Index of any subdivision of AB will always be the same as the orientation of the edge AB, which is the same as the orientation of the original triangle ABC. The proof is by induction on the number of labels that have been used to create the subdivided edge. We have effectively shown above that if we just insert one point of either letter (A or B) our claim is true. Now assume that we are subdividing edge AB and our claim holds for the first n labels that we have inserted. If we insert the label A, it must be between a pair of labels, BB, AB, BA, or AA. Inserting the label A on the segment BB gives rise to BA and AB. They have opposite orientations, so their signed orientations cancel and the Index with the additional point remains unaltered. Inserting the label A on segments with endpoints AB or BA simply creates a new AB or BA, respectively, along with an AA segment. No change. Of course there is no change in the value of the Index if we insert an A on AA. The argument that unfolds when we add the label B is completely analogous.

We can "sum things up" as follows. Anytime we apply a labeling to the edges of a triangle ABC à la Sperner, the Index of the edge AB will coincide with the signed orientation of the triangle, $+1$ or -1. It should be clear that there is nothing special about the edge AB, since edges AC and BC have the same orientation.

As a reminder, our goal is to show that when we triangulate the triangle ABC and do a Sperner labeling, we produce at least one triangle that has a vertex with each of the three labels. Further, we want to show that we will have one more complete triangle with the orientation of the outer triangle than with the opposite orientation. To this end, we introduce another definition. Let the *Content* be the sum of the (signed) orientations of the complete triangles in the triangulation.

If we can show that the Content equals the Index, we're done. Before we take this on, we have just one more definition. Let **S** (for substance) be the number of segments labeled AB in the triangulation counted by adding orientation values ($+1$ or -1), including those on the original triangle. If the segment is in the interior, it is counted twice, once for each of the triangles to which it belongs. We will finish the proof by showing that both the Content and the Index equal **S**.

First, we show that the Content equals S.

What does the complete triangle

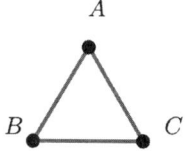

contribute to the Content? Since the orientation is counterclockwise, the answer is $+1$. But this is exactly what it contributes to **S**, since it only has one counterclockwise AB edge.

Now we are on a roll and quickly see that the complete triangle with clockwise orientation also contributes the same quantity (-1) to both **S** and the Content. Triangles that do not have both an A and a B in their labeling contribute zero to **S** and zero to the Content, since they are not complete. Finally, triangles that have two Bs and an A or two As and a B have AB edges with different orientations and contribute zero to **S** and, not being complete, contribute zero to the Content. So, the Content equals **S**!

Show that the Index equals S.

Well, any AB edge in the interior contributes zero to **S**, since it will appear twice with the opposite orientation. It follows that interior edges contribute a net of zero to **S**. That leaves only the exterior AB segments to determine the ultimate value of **S**, whose signed sum also determines the Index. The Index equals **S**!

Suddenly we're done. **Index equals Content!**

We have shown that for any triangle the Content of a Sperner-labeled triangulation can only be -1 or 1, so there must be at least one complete triangle, otherwise the Content would be zero. Further, the triangulation must have one more complete triangle with the same orientation as the original triangle than with the opposite orientation, since the Index equals the signed orientation of the triangle. This whole discussion can be extended beyond triangles to general polygons. A good reference is Michael Henle's book, *A Combinatorial Introduction to Topology.*[1]

$$* * *$$

The Brouwer Fixed-Point Theorem

Sperner's Lemma lays the groundwork for perhaps the most accessible proof that I have seen of the Brouwer fixed-point theorem. Brouwer's theorem tells us that, under a certain condition, if we take a disc (boundary and interior) and assign each point to a point that is also in the disc, at least one point of the disc must be assigned to itself.

We need to say what a disc is. We also have to say what the "certain condition" is. The most natural example of a disc is suggested by the word itself. A circle in the plane along with its interior is an example of a disc. The condition we require for the assignment (transformation, function) is that it be continuous. Imagine the disc being made of fabric. A continuous transformation makes assignments that do not produce a tear in the fabric.

Let's take the circular representation of a disc and try to produce a continuous assignment that assigns every point to a different point. A suggestion from the audience might be to rotate the disc 30 degrees. This transformation certainly does not cause any tears, but it also does not assign every point to a different point. The center of the disc gets assigned to itself.

We can also try to make our understanding of a continuous transformation a bit more mathematical. A tear is caused when "closeness" is not preserved. With a continuous transformation T we can expect that as we take points p_1, p_2, p_3, ... closer and closer to a point p, the assignments $T(p_j)$ of the points p_j get closer and closer to the assignment $T(p)$ of p:

$$\lim_{n \to \infty} T(p_n) = T(p).$$

---

[1] *A Combinatorial Introduction to Topology*, Michael Henle, W.H. Freeman and Company, 1979.

If this doesn't happen we get a tear or puncture in the fabric. Rotating the circular disc and then moving the point at the center someplace else is what I am calling a puncture.

The goal is to prove that for continuous transformations from the disc into itself there is always a point that is not moved, a *fixed point*.

The analogue of a disc on the real line is a closed interval $[a, b]$. The continuous functions from the interval into itself always have a fixed point. The proof for the interval is considerably easier than for the disc. It goes like this.

Let f be a continuous function from the interval $[a, b]$ into itself. To show that there is an x in $[a, b]$ that satisfies $f(x) = x$, we let

$$h(x) = f(x) - x.$$

If $h(a) = 0$ or $h(b) = 0$, then we have a fixed point and are done. Otherwise neither is equal to zero and we must have $f(a) > a$ and $f(b) < b$, since we have ruled out $f(a) = a$ and $f(b) = b$ and the assignments of f must be in $[a, b]$. It follows that $h(a) > 0$ and $h(b) < 0$.

Now we use the continuity of f and, accordingly, the continuity of h to assert that there must be an x in the interval $[a, b]$, with $h(x) = 0$. This is because a continuous function, which h is, cannot take on positive values and negative values in the interval without taking on the value zero. If it did, that would indicate a tear. We have just bypassed something called the *Intermediate Value Theorem* and replaced it with the reader's intuition. We can live with that.

It will not surprise the reader that our selection from the large family of "discs" available is the triangle with its interior, a distortion of the circular disc. We'll use an equilateral triangle with edge-length 1. We intend to take advantage of Sperner's Lemma.

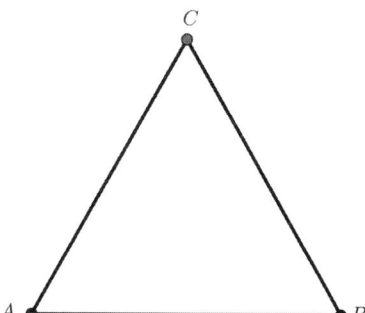

Above is our triangular disc. We assume that we have a continuous transformation T that assigns each point in our disc to a point on the disc. At each point on the disc we place a vector with the tail on the point p and the tip of the vector on the point $T(p)$.

Below, we give an example of what the transformation could do. By
assumption, points cannot be assigned to positions outside the disc. For
convenience, we place our equilateral triangle (edge length 1) in the Cartesian
plane so that the midpoint of the edge AB is at the origin.

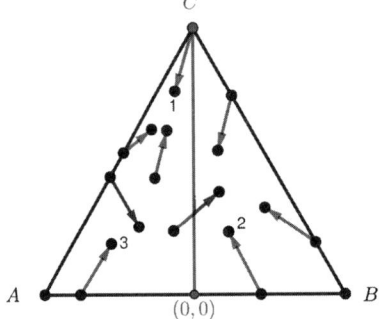

Let the line perpendicular to AB through the origin represent the north-
south axis and the line containing the edge AB represent the east-west axis.
We will label our vectors with three possible directions: northwest, northeast,
and south. We decide on whether a vector is south, northeast, or northwest
in its orientation by placing the tail of the vector at the origin $(0,0)$ and
keeping it parallel to the original vector. We have done that below for the
vectors labeled 1, 2, and 3.

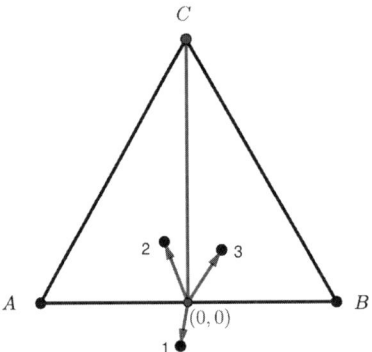

Vector 2, with its head in the second quadrant (to the left of the y-axis
and above the x-axis), is pointing northwest. Similarly, the tip of vector 3
is in the first quadrant and is considered to be pointing to the northeast.
Finally, the area below the horizontal line determined by the edge AB corresponds to the third and fourth quadrants (south). Vector 1 points south.
If a vector is pointing due east it will be viewed as pointing northeast, and
a vector pointing due west will be viewed as pointing northwest. If a vector
is pointing due north, we break the tie and put the northwest label on it.

We are sneaking up on a Sperner labeling convention. If we then say
that all **northwest-pointing** vectors will receive the label B, **northeast-
pointing** vectors the label A, and **south-pointing** vectors the label C, we
see that for any subdivision of AC, for instance, that uses the label of the

vector emanating from the segment, the only possible labels are A or C. See
the graphic below.

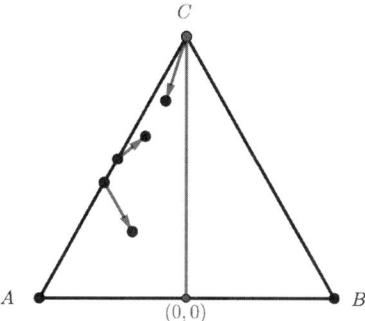

Similarly, edges AB and BC can only receive labels that are consistent
with a Sperner labeling.

Below we picture a triangulation of our equilateral triangle by four equilateral triangles with edge-length $\frac{1}{2}$.

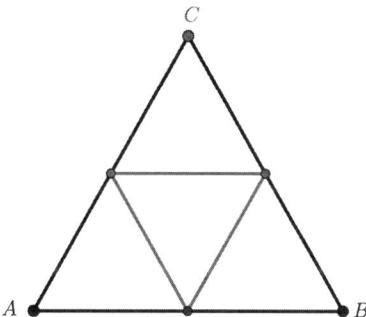

If we follow the recipe for a Sperner labeling that we have just outlined,
each vertex in the triangulation will receive a label and we are guaranteed to
have at least one complete triangle in the triangulation. Remember that it
is the transformation T that determines the direction of the vector at each
vertex, which gives us the vertex label according to our recipe. We pick one of
the complete triangles and label its vertices A_1 (northeast), B_1 (northwest),
and C_1 (south). We subdivide once more and get a finer triangulation. This
triangulation is accomplished by triangulating each of the triangles in the
previous triangulation into triangles with edges of length $\frac{1}{4}$.

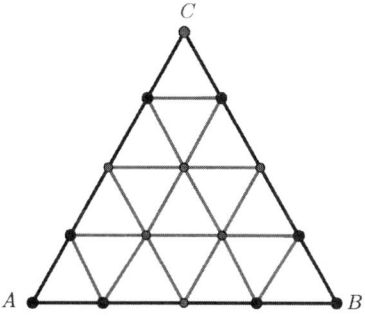

This triangulation also has at least one complete triangulation. We pick one and label its vertices A_2, B_2, C_2. We continue this process ad infinitum, forcing the length of the edges ever closer to zero and producing at each step a set of vertices, A_n, B_n, C_n. We have three sequences, A_n, B_n, and C_n.

Let's go back to the very first in the series of triangulations. For the moment let's just think about the sequence A_n. We only have four triangles in this triangulation. At least one of them must contain infinitely many elements of the sequence A_n.

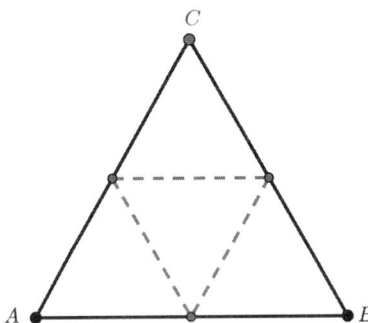

For this discussion, we have outlined a triangle with a broken border that has infinitely many A_n. The second triangulation refines the first. Our outlined triangle will contain four triangles, and one of them must contain infinitely many A_n.

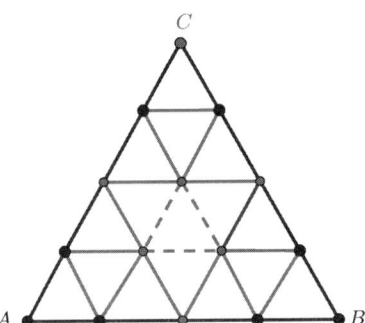

We can clearly keep this going. The broken-edge triangles have edges that are going to zero and the distances between points contained in these triangles goes to zero. Here's a claim that will ease our way through this. (For the reader who is unfamiliar with set notation or a bit rusty, please take advantage of Appendix A as an aid in understanding the next claim.)

Claim: If we have a nested sequence of discs inside a disc, their intersection cannot be empty. Let $\emptyset$ represent the empty set. If we let $\{D_j\}$ be a nested sequence of discs, symbolically we have

$$D_1 \supset D_2 \supset D_3 \supset \ldots, \text{ then } D_1 \cap D_2 \cap D_3 \cap \ldots \neq \emptyset.$$

It is easy enough to modify a disc slightly so that non-empty-nested-intersection property does not hold. If we let each D_n be circular discs with radius $1/n$, all sharing the same center position, but the point at the center itself is removed, then their intersection is empty. But without modifications of this sort the intersection will always be non-empty.

Using the claim, we assert that the intersection of all the broken-edge triangles cannot be empty. It follows that there can only be one point, call it **p**, in the intersection of all these broken-edge triangles. If there were two or more in the intersection, the distance between them would be greater than $1/n$ for some n, which would make it impossible for both of them to be in any triangle with edge length less than $1/n$.

Now we're in a position to knock this out. We have a sequence of A_n vertices getting arbitrarily close to our point **p**. For each A_n there is a B_n and C_n that must be close to **p** also, because they are close to A_n. Of course, the transformation T assigns **p** to only one point and the vector pointing from **p** to that assignment must be very close to the vectors associated with A_n, B_n, and C_n, by continuity. Further, since the vectors associated with A_n, B_n, and C_n are close to each other they must all have lengths close to zero, because they cannot possibly be getting close by aligning in the same direction. We support this last assertion by noting that the angle of rotation counterclockwise from the northeast vector to the south vector must be greater than 90° for each triple. Therefore the tips of the vectors associated with A_n, B_n, and C_n can only get closer to each other if the lengths of the vectors are getting shorter and shorter. As n gets large, the transformation must be making assignments to A_n, B_n, and C_n that are closer and closer to the points themselves, which are getting arbitrarily close to **p**. We conclude that the vector associated with **p** must be the zero vector. Accordingly, the point **p** must be assigned to itself. It is a fixed point. This completes the proof of the Brouwer fixed-point theorem.

Sperner's Lemma generalizes to higher dimensions. For instance, in three dimensions we would use a tetrahedron instead of a triangle and show that the tetrahedron has the fixed-point property. An adaptation of the proof that we have just given even works in the one-dimensional case, the closed interval $[A, B]$. The directions in the interval would just be east and west (right and left). As I have already mentioned, triangles are only one example of a disc. Essentially anything that can be created by continuously stretching a disc, without tearing it, is also a disc. All discs have the fixed-point property. This is true in higher dimensions as well.

I would like to acknowledge again my debt to Michael Henle's book, *A Combinatorial Introduction to Topology*, for all of the essential pieces that I have used in the proof of the Brouwer fixed-point theorem. The minor departures from his fine presentation are due to my idiosyncratic tastes.

A theme that we find repeatedly in mathematics is that computations on the boundary of an object can be surprisingly informative. There is sometimes a closer relationship between a book and its cover than one might expect. Sperner's Lemma is just one case in point. For those of you who have been exposed to the calculus you will recognize another relationship that illustrates this, the Fundamental Theorem of Calculus. It is worth saying this aloud:

Let f be a continuous function. If $f(x)$ measures the slope of the tangent line to the curve described by some function $F(x)$ at each point (x, y) on the curve, then the signed area (it could be negative) of the region bounded by the curve $f(x)$, the x-axis, and the vertical lines $x = a$ and $x = b$ is precisely $F(b) - F(a)$.

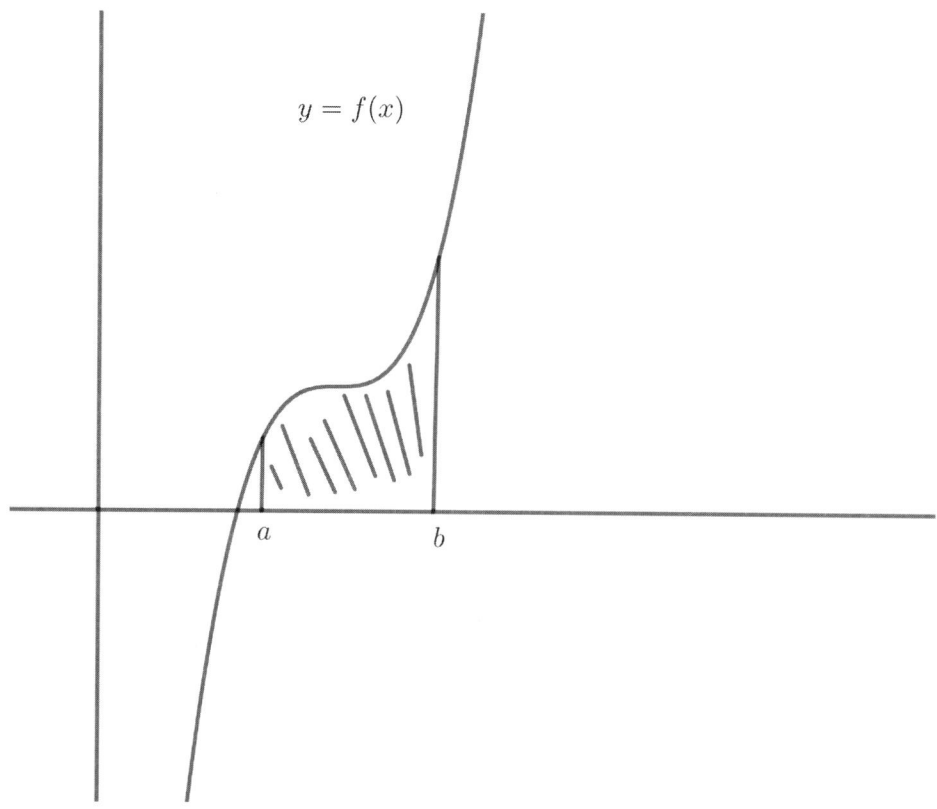

$$\text{Area} = F(b) - F(a)$$

For many functions, including polynomials, F is easy to compute. It is almost as easy as looking at a triangle and determining if ABC (in that order) is counterclockwise or clockwise.

Chapter 12

Consider the Sequence (Fibonacci and Golden Ratio)

Paul Erdős was a twentieth-century mathematician and central player in the mathematics of his times. Both of his siblings, two sisters, were lost to scarlet fever a few days before he was born. Many members of his family, including his father, perished in the Holocaust. His mother survived by going into hiding. He was in the United States at the Institute for Advanced Study in Princeton during that period.

Erdős completed the PhD degree in mathematics at the University of Budapest in 1934. He was twenty-one years old. One of the most prolific mathematicians ever, for a good deal of his life he wandered from campus to campus and conference to conference, staying at the homes of mathematicians, who were simply happy to host Paul Erdős. He had no fixed abode and very little in the way of material possessions. Frequently during his wanderings, he had his mother in tow. He never married and had no children.

Erdős did not believe in God, and said that if God exists he is the Supreme Fascist. Erdős called him SF for short and speculated that SF keeps all of the beautiful and elegant proofs of mathematical theorems for himself in the Book. "You don't have to believe in God, but you should believe in The Book." When a proof is especially elegant, many mathematicians do say that it is from The Book.

There is a concept called the Erdős number. If you published a paper with Erdős, your Erdős number is one. If you have published a paper with someone who has published a paper with Erdős, your Erdős number is two, and so on. My Erdős number is three due to a collaboration with Edray Goins when he was in graduate school at Stanford. Edray did not have an Erdős number at the time, but later picked up Erdős number two. He also provided a proof that summer, when both he and Donald J. Newman were at NSA; a proof that Newman described as being from The Book. But more about that later.

Immediately after completing the PhD, I accepted a position at Auburn University. One of my colleagues there was Curt Lindner, whose specialty is combinatorics. As had happened for many others, Curt received a phone call from Erdős asking if Curt could put him up for a couple of days. Curt expressed a cheerful willingness to host him. This is the story he told me.

> Never again. The man takes enough drugs to kill an elephant. He said that there is enough time to sleep in the grave. Erdős moved non-stop through the house, morning and night. At one point, I woke up and found him standing over the bed. As I opened my eyes he said, "Consider the sequence..." My wife was screaming.

Now that I recall this story, I wish I had told Curt to erect a sign in front of his house: Paul Erdős Never Slept Here.

Erdős contributed to many branches of mathematics. One of them is what is called *Ramsey Theory*. Rather than getting bogged down in definitions, we'll illustrate Ramsey theory with an example. It is sometimes called the "theorem on friends and strangers."

Let's suppose that person A is having a party. How many people must be at the party to guarantee that there are three people present who are total strangers to each other *or* three people who are all friends (each person was already friends with each of the other two)?

Because I already know the answer, we start off with the assumption that there are six people at the party. We are not allowing for a middle ground in this problem, so if no more than two are friends then at least three must be strangers, and vice versa. Let's assume that three of the guests are friends of person A. Call them persons B, C, and D. We depict this below with *solid* line segments of friendship from A to B, C, and D.

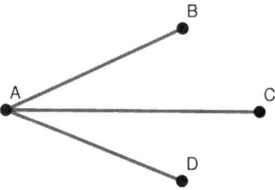

Now, if B and C are friends or C and D are friends, we have a group of three friends. Let's assume that this is not the case. Persons B and C and persons C and D are strangers. Below we indicate that B and C and C and D are strangers by drawing broken line segments between them.

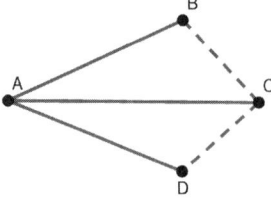

We now note that the relationship between B and D remains undefined. If we say that they are strangers and extend a broken line segment between them we have the following picture.

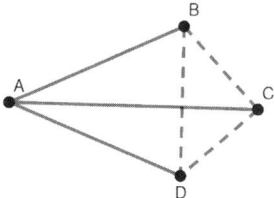

We see that persons B, C, and D are mutual strangers. Well, what if B and D are friends? We have a new formation.

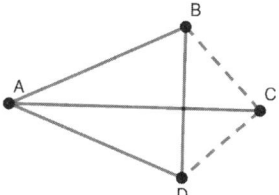

We now see that persons A, B, and D are friends. We have a threesome of friends!

The reader should take a few seconds to establish that we could have begun with broken lines from A to B, C, and D, but we still would not have been able to avoid a threesome of strangers or a threesome of friends.

Suppose that we had a smaller party, say only five. Is it possible to have neither a threesome of friends nor a threesome of strangers? The picture below tells us that the answer is yes.

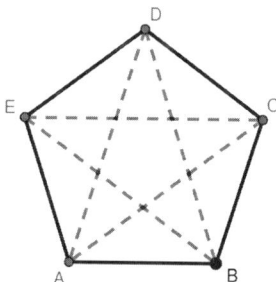

Any triangle in this configuration that has each of its vertices corresponding to a person has either one solid edge and two broken edges or one broken edge and two solid edges. There is no triangle with all broken edges, so there is no threesome of mutual strangers. Similarly, there is no threesome of mutual friends. In Ramsey-speak, one would say that there are no monochromatic cliques. In some presentations two colors are used instead of broken edges and solid edges.

The notation $R(3,3)$ stands for the smallest integer such that we are guaranteed to have a threesome of mutual strangers or a threesome of mutual friends. We have seen that the Ramsey number $R(3,3)$ is 6.

This example may give you the impression that calculating Ramsey numbers is child's play. It is not. In fact, it is surprisingly difficult. Ramsey Theory is named after Frank P. Ramsey, who lived in the early twentieth century and died at the age of 26. It was not proven until more than a half century after Ramsey's death that $R(4,4)$ is 18. At the time of this writing, it is only known that $R(5,5)$ must lie between 43 and 48.

Paul Erdős was without peer. He died at age 83 of a heart attack while attending a mathematics conference in Warsaw. I have heard that he thought a fitting inscription on his gravestone would be, "I have finally stopped getting dumber."[1]

* * *

Now I ask you to consider the sequence:

$$1, 1, 2, 3, 5, 8, 13, 21, \ldots.$$

We have

$a_1 = 1, a_2 = 1, a_3 = a_2 + a_1,$ and generally, for $n > 2, a_{n+1} = a_n + a_{n-1}.$

The sequence is named after the Italian mathematician Fibonacci, because he is believed to be the first to introduce it in Western culture in the 13th century. (Fibonacci is also credited with popularizing Hindu-Arabic numerals in Europe.) The sequence had been studied in India at least fifteen centuries earlier. There is even a journal called the *Fibonacci Quarterly* dedicated to publishing research results related to the sequence. The literature abounds with examples of natural phenomena that are described by the

---

[1] The Erdős biography, *My Brain is Open*, by Bruce Schechter is excellent.

Fibonacci sequence. There are two that I find interesting. The first is the "family tree" of a drone bee. A drone bee is born from an unfertilized egg. He has no father, only a mother. (Female bees have a mother and a father.) If we start counting the bees on the drone's family tree in each generation, starting with the drone himself, we have 1 corresponding to him and 1 corresponding to his mother. He has two grandparents (his mother's parents) and three great grandparents. We see 1, 1, 2, 3, ... In the next chapter we will look at the Fibonacci sequence as it arises in another naturally occurring example involving generations of living organisms. I will even provide a chart.

In this chapter I want to derive the relationship between the Fibonacci sequence and the Golden Ratio. The Golden Ratio, which we will denote with the symbol, ϕ, is

$$\phi = \frac{1 + \sqrt{5}}{2}.$$

We will let f_n denote the nth term in the Fibonacci sequence. There is a nice formula for the nth term. It is attributed to the 19th century mathematician Jacques Binet, even though it was known to Abraham DeMoivre a century earlier. Binet's formula is

$$f_n = \frac{1}{\sqrt{5}} \left(\phi^n - \left(-\frac{1}{\phi} \right)^n \right).$$

A quick computation shows that $\left(-\frac{1}{\phi} \right) \doteq \frac{1 - \sqrt{5}}{2}$. Denote this quantity by $\bar{\phi}$ for future use.

Like the Fibonacci sequence, the Golden Ratio's fame transcends mathematics, its roots entwined in ancient civilizations. Also known as the Divine Proportion, its fingerprints are found in music, art, and biology. In what may be its original incarnation, it describes the ratio that results when you require that a segment be divided into two pieces, a and b, so that the ratio of the whole segment to a is the same as the ratio of a to b.

$$\frac{a+b}{a} = \frac{a}{b}, \text{ so } 1 + \frac{b}{a} = \frac{a}{b}.$$

We are interested in the ratio a/b. Call it x. Then

$$1 + \frac{1}{x} = x, \text{ so } x^2 - x - 1 = 0.$$

The quadratic formula provides two solutions, ϕ and $\bar{\phi}$. We choose the positive one, ϕ.

The fact that the two mathematical objects, ϕ and f_n, enjoy the cozy relationship described by Binet's Formula is yet another wonder.

I am going to take the long way around to get to Binet's Formula. One might call it the scenic route. There are faster vehicles, but speed is not the issue for us.

Here is a problem that we learned to do in grade school. Given a number with a decimal expansion that is repeating, represent it as the ratio of two integers.

$$x = .7371737173717371\ldots,$$
$$10000x = 7371.737173717371\ldots.$$

Subtracting, we get

$$9999x = 7371$$

and

$$x = \frac{7371}{9999} = \frac{819}{1111}.$$

Now, just for fun, form an infinite series with the Fibonacci numbers as the coefficients:

$$F(x) = x + x^2 + 2x^3 + 3x^4 + 5x^5 + 8x^6 + 13x^7 \ldots.$$

When we display the values of a sequence this way, the series is called a *generating function*. Inspired by the grade school arithmetic we just dredged up, we play:

$$F(x) = x + x^2 + 2x^3 + 3x^4 + 5x^5 + 8x^6 + 13x^7 \ldots,$$
$$xF(x) = x^2 + x^3 + 2x^4 + 3x^5 + 5x^6 + 8x^7 \ldots,$$
$$x^2 F(x) = x^3 + x^4 + 2x^5 + 3x^6 + 5x^7 \ldots.$$

Subtracting the two bottom lines from the top one, we get *a lot* of cancellation and have

$$\left(1 - x - x^2\right) F(x) = x.$$

The reader may be feeling uncomfortable at this point. How do we know that these operations are justified? How do we know that these series add up? We shall forge ahead formally without answering these questions and rescue ourselves at the end of the tale. So, solving for $F(x)$, we have

$$F(x) = \frac{x}{1 - x - x^2}.$$

So what?

Well, if we do a little algebra we see that

$$F(x) = \frac{x}{1 - x - x^2} = \frac{1}{\sqrt{5}} \left(\frac{1}{1 - \frac{(1+\sqrt{5})x}{2}} - \frac{1}{1 - \frac{(1-\sqrt{5})x}{2}} \right).$$

You can take this on faith or go ahead and simplify the rightmost expression to verify that this equality holds. If you have been exposed to the method of partial fractions, perhaps in calculus, you will know how I managed to come up with this.

Remaining oblivious to whether the series that I've introduced are summable or not, I remind the reader that

$$\frac{1}{1-u} = \sum_{k=0}^{\infty} u^k.$$

Of course, there is the little matter of

$$|u| < 1$$

that we are not paying close attention to at the moment. If we apply this formula for the sum of a geometric series to the expressions

$$\left(\frac{1}{1 - \frac{(1+\sqrt{5})x}{2}} - \frac{1}{1 - \frac{(1-\sqrt{5})x}{2}} \right),$$

we get

$$F(x) = \frac{x}{1 - x - x^2} = \frac{1}{\sqrt{5}} \left(\frac{1}{1 - \frac{(1+\sqrt{5})x}{2}} - \frac{1}{1 - \frac{(1-\sqrt{5})x}{2}} \right)$$

$$= \frac{1}{\sqrt{5}} \sum_{n=0}^{\infty} \left(\frac{1+\sqrt{5}}{2} x \right)^n - \left(\frac{1-\sqrt{5}}{2} x \right)^n.$$

So, we see that the coefficient of x^n, supposedly the nth Fibonacci number, is

$$\frac{1}{\sqrt{5}} \left(\left(\frac{1+\sqrt{5}}{2} \right)^n - \left(\frac{1-\sqrt{5}}{2} \right)^n \right) = \frac{1}{\sqrt{5}} \left(\phi^n - \bar{\phi}^n \right).$$

Binet's Formula!

What about the summability question? It is easy enough to check that

$$\left| \left(\frac{1+\sqrt{5}}{2} x \right) \right| < 1, \text{ if } |x| < \frac{\sqrt{5}-1}{2}.$$

It is also true that

$$\left| \left(\frac{1-\sqrt{5}}{2} x \right) \right| < 1, \text{ if } |x| < \frac{\sqrt{5}-1}{2}.$$

So if we wanted to evaluate the series that corresponds to the generating function F at, say, $x = \frac{1}{2}$, the following equality does hold:

$$2 = \frac{\frac{1}{2}}{1 - \frac{1}{2} - \left(\frac{1}{2} \right)^2}$$

$$= \frac{1}{2} + \left(\frac{1}{2} \right)^2 + 2 \left(\frac{1}{2} \right)^3 + 3 \left(\frac{1}{2} \right)^4 + 5 \left(\frac{1}{2} \right)^5 + 8 \left(\frac{1}{2} \right)^6 + 13 \left(\frac{1}{2} \right)^7 + \ldots$$

That's because $\frac{1}{2}$ lies inside the interval where the series is summable. However, this is not essential to the argument we were trying to make.

We wanted a formula for the nth Fibonacci number and we got it. The approach we took involved using a generating function. Since in this case we already had a candidate for the formula, it would not have been hard to verify directly that the formula works. If we evaluate Binet's Formula at $n = 1$, we get 1. At $n = 2$ we also get 1. Now all we need to do to wrap things up is show that

$$\frac{1}{\sqrt{5}}\left(\left(\frac{1+\sqrt{5}}{2}\right)^{n-1} - \left(\frac{1-\sqrt{5}}{2}\right)^{n-1}\right) + \frac{1}{\sqrt{5}}\left(\left(\frac{1+\sqrt{5}}{2}\right)^{n} - \left(\frac{1-\sqrt{5}}{2}\right)^{n}\right)$$
$$= \frac{1}{\sqrt{5}}\left(\left(\frac{1+\sqrt{5}}{2}\right)^{n+1} - \left(\frac{1-\sqrt{5}}{2}\right)^{n+1}\right).$$

This looks like a mess, but we've already been in worse spots in this book, believe me. We simply have to recall that both ϕ and $\bar{\phi}$ satisfy

$$x^2 - x - 1 = 0.$$

So $\phi + 1 = \phi^2$. Hence

$$\phi^n + \phi^{n-1} = \phi^{n-1}\left(\phi + 1\right) = \phi^{n-1}\phi^2 = \phi^{n+1}$$

and

$$\bar{\phi}^n + \bar{\phi}^{n-1} = \bar{\phi}^{n-1}\left(\bar{\phi} + 1\right) = \bar{\phi}^{n-1}\bar{\phi}^2 = \bar{\phi}^{n+1}.$$

And that settles it. We have just done what I mentioned above. We have simply verified Binet's formula by induction. Why did we bother with that long trip around the barn? The answer: The ideas we examined are generally useful. Further, you may encounter situations where the only way to truth is around the barn.

Just by the way, since $|\frac{1}{\sqrt{5}}\bar{\phi}^n| < \frac{1}{2}$, Binet's Formula tells us that rounding $|\frac{1}{\sqrt{5}}\phi^n|$ to the nearest integer produces the nth Fibonacci number.

What about the ratio of successive elements of the Fibonacci sequence?

$$\lim_{n\to\infty} \frac{f_{n+1}}{f_n} = \lim_{n\to\infty} \frac{\frac{1}{\sqrt{5}}\left(\phi^{n+1} - \bar{\phi}^{n+1}\right)}{\frac{1}{\sqrt{5}}\left(\phi^n - \bar{\phi}^n\right)}$$
$$= \lim_{n\to\infty} \frac{\left(\phi^{n+1} - \bar{\phi}^{n+1}\right)}{\left(\phi^n - \bar{\phi}^n\right)}$$
$$= \lim_{n\to\infty} \frac{\phi - \frac{\bar{\phi}^{n+1}}{\phi^n}}{1 - \frac{\bar{\phi}^n}{\phi^n}} = \phi.$$

The last equality holds because

$$|\bar{\phi}| < 1 \text{ and } \phi > 1, \text{ so } \lim_{n\to\infty}\frac{\bar{\phi}^n}{\phi^n} = \lim_{n\to\infty}\frac{\bar{\phi}^{n+1}}{\phi^n} = 0.$$

Here's a string of equalities for the reader to play with and discover why they hold:

$$\phi = \sqrt{1 + \sqrt{1 + \sqrt{1 + \sqrt{1 + \sqrt{1 + \ldots}}}}} = 1 + \cfrac{1}{1 + \cfrac{1}{1 + \cfrac{1}{1 + \cfrac{1}{1 + \ldots}}}}.$$

It is six in the morning
The house is asleep
Nice music is playing
I prove and conjecture...
——Paul Erdős for Vera Sós–

Chapter 13

What Are the Chances?

I've always viewed probability as tricky business. Intuition frequently fails to be a reliable guide. It is not unusual for me to think that two irreconcilable claims are both right, which is, of course, upsetting. It is relatively easy to demonstrate how quickly things can go off the rails. The problem is called the Monty Hall problem and the reference is to the television game show, *Let's Make a Deal*, which was hosted by Monty Hall. Now that I have described the origin of the name, I'm going to move straight to a description of the problem without any further history of the game show. Here is a statement of the question stripped of many of the trappings.

Assume that you are in a competition with an opponent for a prize. The prize is hidden behind one of three doors. For whatever reason, your opponent is given an advantage. You are only allowed to pick one of the doors. The two doors that you do not pick are then assigned to your opponent. There is an additional twist. Your opponent is asked to peek behind each of her doors and open one door, but with the stipulation that she must not open a door that has the prize behind it. Once your opponent carries out this action, you are given the option of choosing the door that she did not open or staying with the door you were originally assigned. If you want to maximize your chance of winning, what decision do you make?

This question gave rise to a bit of controversy in the early nineties, with many people weighing in on what the correct decision would be. Some believed that switching made no difference. Others insisted that it did. This is the only case that I can recall of a mathematical question receiving such widespread popular coverage.

While the reader lets this rattle around in the subconscious, we'll do some simple exercises in probability. We have compiled some information on family and friends in the following table. We let PAT and MAT stand for paternal relatives and maternal relatives, respectively. We also assume that we are not related to the eleven friends in our "study" and that no paternal relative is also a maternal relative. We have a total of twenty-four different people that we have examined to determine if they are left-handed or not.

	Left-handed	Not Left-handed	Total
PAT	3	2	5
MAT	1	7	8
Friends	2	9	11
Total	6	18	24

If I were to ask what the probability is that I get a left-handed person, if I randomly choose a person from these twenty-four, my guess is that almost all readers will say that it is the number of left-handed people among our twenty-four divided by twenty-four:

$$\frac{6}{24}.$$

Of course, we can ask further questions. Again, choosing randomly from the twenty-four people in our collection, what is the probability of choosing a person who is both a friend and not left-handed. Well, there is a box that tells us that the number of people who are both friends and not left-handed is nine. So the probability that we get such a person in a random selection is

$$\frac{9}{24}.$$

Let's ask a more "refined" question. Given that you know you have picked a person who is a paternal ancestor, what is the probability that the person is left-handed? We have only five paternal relatives and three are left-handed, so given that we know we have selected from that group of relatives, the answer is

$$\frac{3}{5}.$$

It is useful to look at answering this question another way. Note that

$$\frac{3}{5} = \frac{\frac{3}{24}}{\frac{5}{24}} = \frac{\text{Probability(left-handed and paternal)}}{\text{Probability(paternal)}}.$$

We are effectively saying that

$$\text{Probability(left-handed, given paternal)}$$
$$= \frac{\text{Probability(left-handed and paternal)}}{\text{Probability(paternal)}}.$$

If we substitute the symbol $\cap$ for the word "and" along with substituting the symbol "$|$" for the word "given," we have

$$\text{Probability(left-handed|paternal)} = \frac{\text{Probability(left-handed} \cap \text{paternal)}}{\text{Probability(paternal)}}.$$

We have effectively given the definition of conditional probability: The probability of B, given that we know that A occurred, can be reduced to the right-hand side of the equality below. We have substituted the letter "P" for "Probability." (We remark here that $P(A) \neq 0$, because we are assuming that A occurs.)

$$(*) \qquad P(B|A) = \frac{P(A \cap B)}{P(A)}.$$

Why bother with writing our definition in terms of probabilities? We had arrived at the answer $\frac{3}{5}$ very easily without resorting to such convoluted reasoning. Well, it turns out that this definition is useful because in many, if not most, applications we only know the probabilities. If we consider the Monty Hall problem we know that the probability of the prize being behind a randomly selected door is $\frac{1}{3}$. We do not have a table with little boxes in this problem. Before we return to the Monty Hall problem, and clear it up completely, we would be well served to take another look at the definition $(*)$.

It follows from the symmetry between A and B that

$$P(A|B) = \frac{P(B \cap A)}{P(B)}.$$

Then

$$P(B \cap A) = P(A|B)P(B).$$

Since "A and B" is the same as "B and A" in our mathematical universe, we can now rewrite the definition $(*)$ to get

$$P(B|A) = \frac{P(A|B)P(B)}{P(A)}.$$

This equality is the most basic form of what is called *Bayes' Theorem*. This appears to be even more unwieldy. In general, it really isn't, but what is more important is that Bayes' Theorem is very useful.

Now let's take the Monty Hall problem apart. When explaining it to friends, I first try to do it with language, not formalism. The way things are set up initially your opponent has probability $\frac{2}{3}$ of having the prize, because she has two doors and you only have one. However, the moderator insists that your opponent show what is behind one of her doors, but stipulates that she cannot open a door that has the prize behind it. Here's where I say, "She opens the door and reveals that there is nothing behind one of her doors. But you already knew that! This does not change the fact that the probability that she has the prize is $\frac{2}{3}$. Now you know that the probability that the prize is behind the door that she didn't open is $\frac{2}{3}$. The moderator's requirement has shifted the advantage to you. Take the door that she did not open and you double your chances of getting the prize." Unfortunately, this spoken argument frequently does not work for the person listening to me. Here is the formal argument using Bayes' Theorem.

We let B be the event that you have the prize behind your door and A be the event that your opponent opens a door that does not have the prize behind it.

Then, as observed above,

$$P(B|A) = \frac{P(A|B)P(B)}{P(A)}.$$

But

$$P(A|B) = P(A) = 1.$$

This is true because your opponent is going to open a door that does not display the prize, whether you have it or not. We knew at the outset that the probability that you have the prize is $\frac{1}{3}$:

$$P(B) = \frac{1}{3}.$$

That leaves us with the probability that you have it being calculated in this way:

$$P(B|A) = \frac{P(A|B)P(B)}{P(A)} = \frac{1}{3}.$$

You conclude that the probability is $\frac{2}{3}$ that the prize is behind the door that your opponent did not open and, for that reason, you throw your fate behind your opponent's unopened door.

That was a lot of work for what we like to think we could have answered in a matter of seconds. We could have simply used the power of the language that we have at our disposal to say that your opponent's chance of having the prize is not reduced by her looking behind both of her doors and opening one that does not have the prize behind it.

What if your opponent is required to open one of her two doors randomly without peeking? Suppose further that when she does this the prize is not behind it. You are again given the option to choose the door that she did not open or stay with the one you have. What now?

We'll use Bayes' Theorem again. Let A represent the prize not being behind the door she chooses to open. Let B represent the prize being behind your door. We claim that

$$P(A|B) = 1,$$

since the prize being behind your door assures that it cannot be behind any other door.

The prize is behind any one of the doors with equal probability and there are three doors. So

$$P(B) = \frac{1}{3}.$$

Using analogous reasoning,

$$P(A) = \frac{2}{3}.$$

So

$$P(B|A) = \frac{P(A|B) \cdot P(B)}{P(A)} = \frac{1 \cdot \frac{1}{3}}{\frac{2}{3}} = \frac{1}{2}.$$

We see in this scenario that it doesn't matter if you switch or not. It's 50-50.

In summary, whether or not your opponent gets to peek first before opening one of her doors and not revealing the prize, you cannot hurt yourself by switching to her door. You certainly might reduce your chances by not switching.

* * *

I mentioned in the previous chapter that we would revisit the Fibonacci sequence and I make good on my promise now. (Eventually I'll tie this in with Bayes' Theorem.) We talked briefly about the fact that the number of ancestors for each generation on the family tree of the drone bee corresponds to the Fibonacci sequence. The same thing holds true for the ancestors for the human male who contribute to his X chromosome.

This is because the human male receives an X chromosome only from his mother. Women receive an X chromosome from both of their parents. A graphic might help (circles female, squares male). The X-chromosome contributors are in gray.

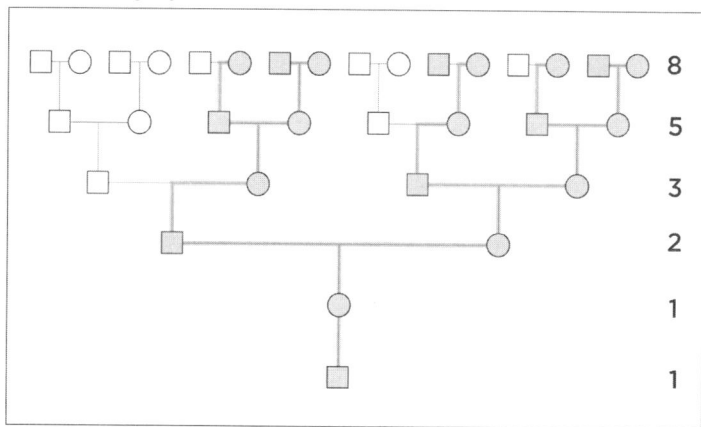

It always feels like an IQ test to me, when I try to talk my way through showing that this is a Fibonacci inheritance pattern, but here goes. First, I ask myself how many X-chromosome contributing ancestors do I have in the $(n + 1)^{\text{th}}$ level of my family tree, written in terms of the number of X-chromosome contributing ancestors on the nth level of the tree. Well, for each female at the nth level I must have one female and one male ancestor at the $(n + 1)^{\text{th}}$ level. For each male ancestor at the nth level I must have one female ancestor at the $(n + 1)^{\text{th}}$ level.

So the number of X-chromosome contributing ancestors, A_{n+1}, at the $(n + 1)^{\text{th}}$ level is given by the sum of twice the number of female contributing ancestors, F_n, at the nth level, plus the number of male contributing ancestors, M_n, at the nth level:

$$A_{n+1} = 2F_n + M_n.$$

On the other hand, if we let A_n and A_{n-1} denote the X-chromosome contributing ancestors at the nth and $(n-1)^{\text{th}}$ levels, respectively, we have

$$A_{n+1} = 2F_n + M_n = F_n + (F_n + M_n) = A_{n-1} + A_n.$$

We see then that the Fibonacci recursion, which was described when we introduced this sequence, is satisfied.

I am an avid genealogist. Since the advent of direct-to-the-consumer DNA testing I have taken advantage of various analyses to help solidify what I know about my anthropological origins as well as my ancestry within genealogical time. (I'm going to view ancestors living within genealogical time as roughly those ancestors who lived no more than 500 years before I was born.) My ethnicity is African American. And as is almost always the case for members of my ethnic group, I have biogeographical contributions within the past five hundred years from more than one continent. In fact, in my case three different regions have impacted my genome, West Africa, Western Europe, and North America. The analysis of my X chromosome was particularly interesting, since I received a highly disproportionate contribution from indigenous North America. The contribution to my X chromosome from a Native American ancestor(s) covered forty-three percent of the chromosome. The overall contribution to my genome from North America is around three or four percent.

Two questions might come to mind. How would this be determined? Why is the X chromosome so out of whack with the overall contribution? The second question is easier to answer than the first. Looking at the chart, we see that of sixteen great-great-grandparents, only five of them are on the "glide path" to my X chromosome. Suppose that the Native American ancestor who made the contribution was a great-great-great-grandmother. She would have been one of eight (sixth Fibonacci number) from that generation who could have contributed to my X chromosome. We see below a possible path (bold line) from her to me. As in the previous chart, we do not display my paternal ancestors.

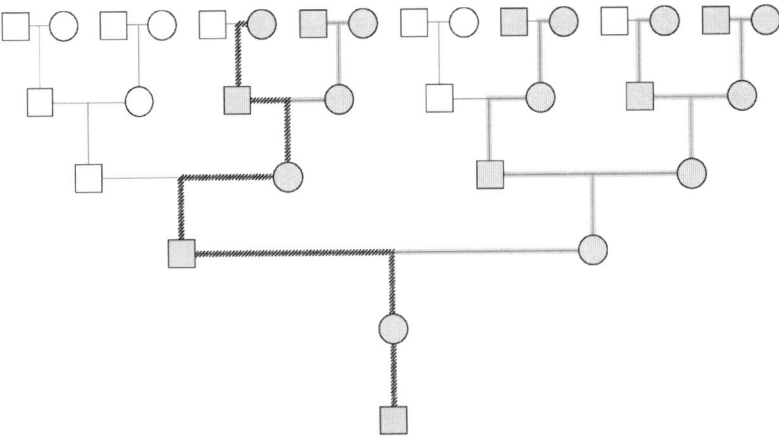

On the way from her to me there are only two recombination events, since the males pass the X to their daughters virtually unchanged. If this were the scenario, it is not at all surprising that a large contribution from this great-great-great-grandmother remained intact all the way from her, a woman who was likely born in the 1700s, to me.

Below is a graphical display of my X chromosome. Note that the ends of the chromosome are light gray, which indicates that they were contributed by West African ancestors. The central section of the chromosome is dark gray, indicating a contribution from the indigenous North American population.

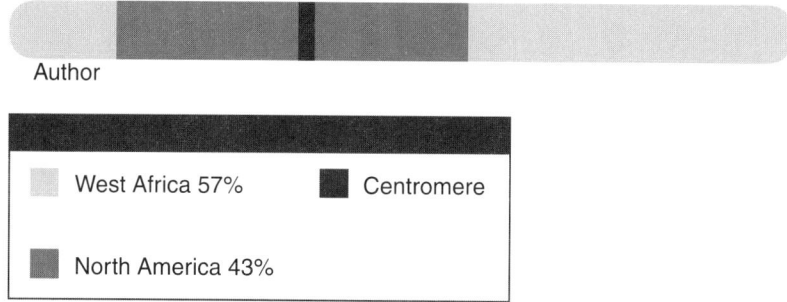

There were no data for the region called the centromere. It is likely in this case that it is also a contribution from the indigenous American biogeographical region, since that region's contribution is on both sides of it.

Now on to the second question. How do we know that the segment on the X chromosome came from a Native American ancestor? The key to answering this question lies in Bayes' Theorem,

$$P(B|A) = \frac{P(A|B)P(B)}{P(A)}.$$

Mutations in the DNA sequence, which occur throughout the human population, often occur at different frequencies in different biogeographical regions. We see that Bayes' Theorem describes a relationship between $P(B|A)$ and $P(A|B)$. What we can establish by looking at various human populations is the probability of a mutation for a given population. What we want to do to establish ancestral contributions is find the probability of a donor population given the observed mutations. Leaving the matter at this would put too simple a face on it. To take full advantage of Bayes' Theorem it is necessary to embed it in a very sophisticated methodology called Hidden Markov Modeling (HMM). HMM was developed around 1970 and one of its original applications was to the analysis of communications, signal processing.

I have written a couple of papers on using HMM to do ancestry analysis. While this book is clearly not the place to delve into this topic any further, it should be said that this material is not inaccessible to a strong and motivated high school student. I had the pleasure of guiding such a student through this material during a year-long mentor/intern project. Xiaolu was a high school junior. She completely digested the methodology. Then she wrote code to implement the model and replicate my results. She was my last intern before retirement. At the time of this writing, Xiaolu is completing her freshman year at MIT.

Chapter 14

The Euler Line

The Euler line is one of the most fascinating objects that I have encountered in geometry. It has the feel of doing astronomy in the Cartesian plane. The picture of an equilateral triangle below will launch us into the adventure.

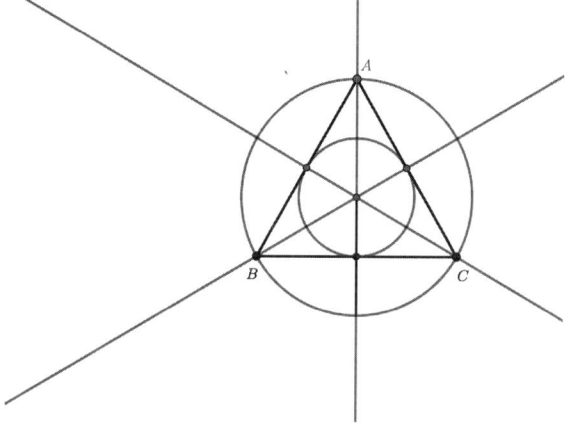

Besides the triangle itself, a few other features are displayed: the *circumcircle*, which passes through the vertices A, B, C, as well as the three altitudes. Formally, an altitude is the perpendicular line segment from the vertex to the line containing the edge of the opposing side. But we will use the term more loosely to include the line containing this line segment. Think of this line as the *extended* altitude. The extended altitudes are the objects that are guaranteed to have a point in common.

The smaller circle is the *incircle*. The incircle is a circle that touches all three sides but does not cross any of them. Note that in the case of the equilateral triangle, it includes the midpoint of each edge. When we move the discussion to arbitrary triangles, intersections with all three midpoints of the edges will be the important aspect, and we will then refer to the circle that passes through all three midpoints as the *nine-point circle*, whether it crosses the edge or not.

We should mention that the altitudes of an equilateral triangle are also the perpendicular bisectors of the edges. Again, it is only in the case of equilateral triangles that the perpendicular bisectors and the altitudes are the same. We see that the perpendicular bisectors intersect at a common point and that point, as it turns out, is also the center of the circumcircle. In some sense we might think of this point as the "center" of the triangle.

Let's see what happens in the general case. Below is a triangle that is clearly not equilateral. We have three line segments, each drawn from a vertex to the midpoint of the opposite side. These are called *medians*. The segments have a point in common, just as the altitudes of the equilateral triangle do, although none of the segments is perpendicular to the edge that it intersects in this case. The point where the three medians intersect is called the *centroid*. Don't worry. Eventually we are going to show that the medians really do intersect at a common point.

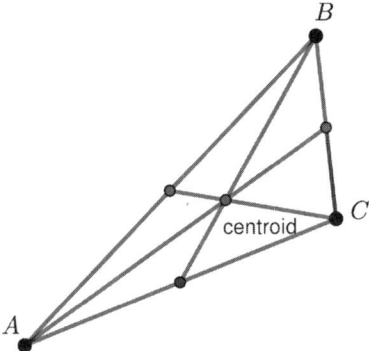

Let's consider the perpendicular bisectors of the triangle, which we display below. The three perpendicular bisectors have a point in common and it is called the *circumcenter*. Yes, it is the center of the circle that passes through the three vertices, the circumcircle. Later in this chapter we will explain why this is true.

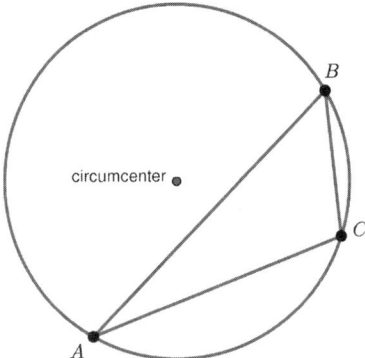

Unlike the case of the equilateral triangle, the circumcenter of this triangle is not even in the interior of the triangle.

We are stubborn and persist. What about the three altitudes? Their common point is called the *orthocenter*. Below we display the orthocenter for our triangle.

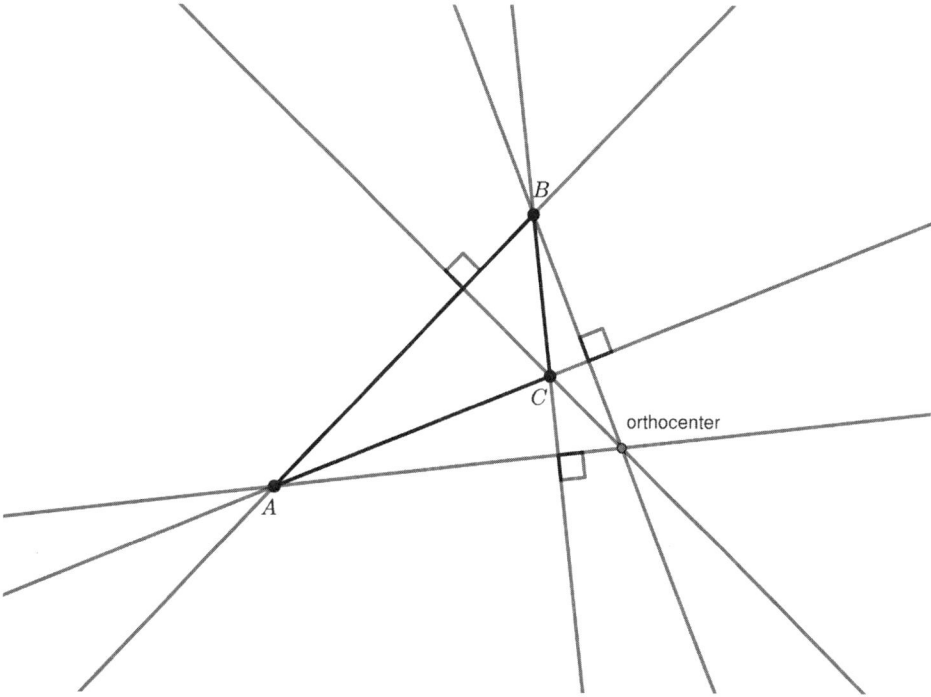

Finally, we consider what is called the *nine-point circle*. It passes through the midpoints of the triangle's edges. We will call its center the *nine-point center* and explain the name in the next paragraph.

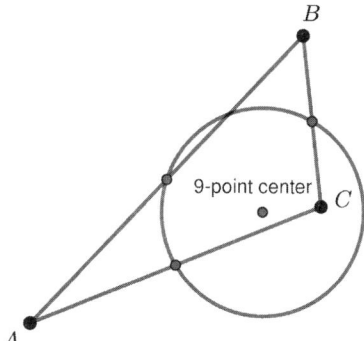

The nine-point circle passes through the midpoints of the three edges, the feet of the three altitudes, and bisects the line segments between the vertices and the orthocenter. That's nine important points!

The four points, centroid, circumcenter, orthocenter, and nine-point center, coincide for an equilateral triangle only, but they do have an amazing relationship in general. They are all on the same line! This line is called the *Euler line*.

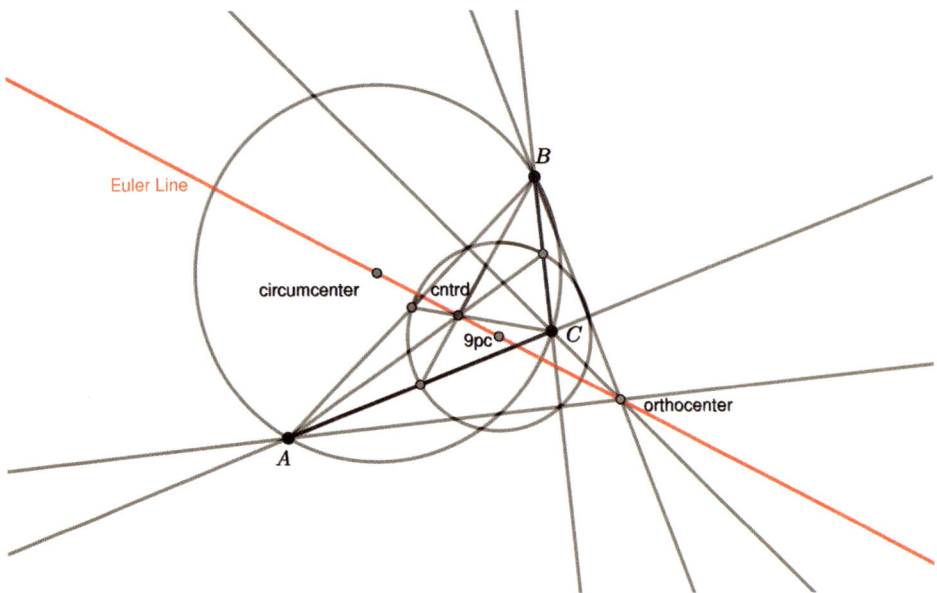

The "centers" always appear in this order on the Euler line: orthocenter, nine-point center, centroid, circumcenter. The nine-point center is the midpoint of the line segment that connects the orthocenter and the circumcenter. The centroid is always two-thirds of the way from the orthocenter to the circumcenter. The radius of the circumcircle is twice the radius of the nine-point circle.

* * *

Now that we have these fantastic claims in front of us, it would be nice to verify that they are true. But first a little bit about the mathematician Euler. Leonhard Euler lived from 1707 to 1783. He was born in Switzerland and held professorships in St. Petersburg and Berlin. Two constants are named after him, the Euler-Mascheroni constant and the base of the natural logarithm, $e = 2.71828\ldots$, Euler's number. Euler could well be viewed as the most productive mathematician in history. In 1775 alone, he authored over fifty papers at a time when he was effectively blind. Euler worked his wonders in other fields including Physics, Astronomy, and Music. He and his wife Katharina brought thirteen children into the world. Five of them lived to adulthood. Euler discovered the Euler line in 1765.

Euler made seminal contributions to an area of mathematics called *complex analysis*. At the heart of complex analysis is the square root of -1, which mathematicians have given a simple name. It is called i.

$$i^2 = -1.$$

The concept first emerged in the 16th century during the quest to find formulas for the roots of polynomials. (This period in mathematics was discussed in the *Impossibilities* chapter.) Complex Analysis has applications that are found throughout mathematics, science, and engineering.

Accepting the square root of -1 was not a natural matter for me. I doubt, if given many chances to discover it, that I would have ever found my way to the concept on my own. Somehow, I managed to get myself to walk through the door that complex analysis opened. I did this with the hope that a deeper understanding would eventually come. One of the first things we encounter in complex analysis is Euler's formula,

$$e^{i\theta} = \cos\theta + i\sin\theta.$$

By letting $\theta = \pi$ this leads immediately to the one of the most aesthetically pleasing equations in mathematics,

$$e^{i\pi} + 1 = 0.$$

I do not have anything to say about it that has not already been said. There is probably no other relationship that captures so much of mathematics so compactly. When so many apparently disparate pieces come together in this way, we are encouraged to look for additional signs. Here is an intriguing and immediate follow-on to Euler's identity:

$$e^{i\pi} = -1,$$

$$\left(e^{i\pi}\right)^{\frac{1}{2}} = (-1)^{\frac{1}{2}} = i,$$

$$i = e^{\frac{i\pi}{2}}.$$

So

$$i^i = \left(e^{\frac{i\pi}{2}}\right)^i = e^{\frac{-\pi}{2}} = \frac{1}{e^{\frac{\pi}{2}}}.$$

This is really strange, but that is what happens sometimes when you walk the walk. We can write it in a slightly different way to drive the point home.

$$\sqrt{-1}^{\sqrt{-1}} = \frac{1}{e^{\frac{\pi}{2}}} = 0.2078795\ldots.$$

$$* * *$$

Let's get on with the business of showing that the orthocenter, nine-point center, centroid, and circumcenter are collinear. One day the reader may see the elegant proof covering much of this using *Sylvester's Triangle Problem* and vectors. The proof that I outline below is quite simply a brute force approach. It has the benefit of providing a formula for the Cartesian coordinates of each of the four centers and the radii of the two circles. Further, it gives a formula for the slope of the Euler line. The "elegant" approach does not provide this information and the proof does not include the nine-point circle.

This is what we will do. We can assume (and so we will) that one of the points, namely C, is at the origin $(0,0)$ of the Cartesian plane. We do this to make the computations easier. The points A and B will be described by the coordinates (a_1, a_2) and (b_1, b_2). Given the vertices of the triangle $C = (0,0)$, $B = (b_1, b_2)$, and $A = (a_1, a_2)$, we compute the coordinates of the orthocenter, nine-point center, centroid, and circumcenter in terms of b_1, b_2, a_1, and a_2. In the process of doing this, we confirm that the three perpendicular bisectors have one point in common. We confirm this for the three medians and the three altitudes also. To verify that the four points are collinear, we then show that the slope of the line through the centroid and each of the other three centers is the same.

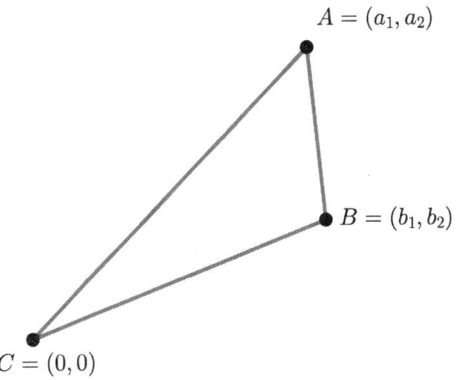

The centroid is the easiest of the four. We now show that the medians meet at the point that lies two-thirds of the way between each vertex and the midpoint of the opposite side. The midpoint of edge AB is simply:

$$V_{AB} = \left(\frac{a_1 + b_1}{2}, \frac{a_2 + b_2}{2} \right).$$

The midpoints of edges AC and BC are

$$V_{AC} = \left(\frac{a_1}{2}, \frac{a_2}{2} \right) \text{ and } V_{BC} = \left(\frac{b_1}{2}, \frac{b_2}{2} \right),$$

respectively. Using the vector arithmetic that we introduce in Appendix A and also discuss in Chapter 3, we find that the point that is two-thirds of the way between $C = (0,0)$ and the midpoint of edge AB is at the tip of two-thirds of the vector V_{AB}:

$$\frac{2}{3} V_{AB} = \left(\frac{a_1 + b_1}{3}, \frac{a_2 + b_2}{3} \right).$$

We verify that two-thirds of the way along the median from vertex A to the midpoint of the opposite edge gives us the same point by adding an appropriate vector to A:

$$A + \frac{2}{3} \left(V_{BC} - A \right) = (a_1, a_2) + \frac{2}{3} \left(\left(\frac{b_1}{2}, \frac{b_2}{2} \right) - (a_1, a_2) \right) = \left(\frac{a_1 + b_1}{3}, \frac{a_2 + b_2}{3} \right).$$

The computation for the remaining case, vertex B, is completely analogous.

It's a much harder slog for the remaining three centers and I will not provide the level of detail that we just went through with the centroid, lest the nursery rhyme become a grim fairy tale. I will confess at this point that, after setting things up, I did not do all of the simplifications by hand, although I did some of them. My investment in the Mathematica software package paid off.

Center, (x, y), and Radius of the Circumcircle

The circumcircle passes through A, B, and C. Its center is also the point of intersection of the lines corresponding to the three perpendicular bisectors of the edges. We outline how to establish this by first carrying out the computations to find the point that the perpendicular bisectors have in common and then we show that the common point is the same distance from each vertex, hence the center of the circumcircle.

We will need the midpoints of the edges, but we already know these points from the work we did in computing the centroid. To find the equations of the lines perpendicular to those midpoints, we need to know the slopes. Establishing the slopes turns out not to be hard. The slopes of the lines that contain the edges AB, AC, and BC are

$$\frac{a_2 - b_2}{a_1 - b_1}, \quad \frac{a_2}{a_1}, \quad \frac{b_2}{b_1},$$

respectively. So the slopes of the perpendicular bisectors of these edges are the negative reciprocals:

$(*)$ $$\frac{b_1 - a_1}{a_2 - b_2}, \quad -\frac{a_1}{a_2}, \quad -\frac{b_1}{b_2}.$$

The alert reader might ask, "What if a_1 or b_1 is zero?" The same alert reader will have observed that not both of them can be, since that would mean that all three vertices are on the y-axis and we would not have a triangle. If a_1 is zero, then the edge from $(0,0)$ to A is on the y-axis, so the perpendicular bisector of that edge would be a horizontal line. Its slope would then be zero. We make the same observation for b_1 and conclude that the information on line $(*)$ above gives us the correct result.

Of course, there are other nagging questions. "What if a_2 equals b_2? What if a_2 or b_2 equals zero?" There are two ways to tackle these issues. The first would be to notice that in all of these cases the perpendicular bisector of the impacted edge will be a vertical line. The equation for this vertical line is easy to describe. However, when you go through the computations for these special cases, you will discover that you could have gotten the same thing by assuming that the nasty conditions do not hold and proceeding as we do below. There is another way to handle these questions. To answer the first question look at the behavior of the general formulas as a_2 approaches b_2. For the second question look at the behavior of the formulas as either a_2

or b_2 approaches zero. Either way you take it on, you will see that you get the same result for these special cases.

Returning to our general attack on the problem, now that we have a point and the slope for each perpendicular bisector, we can write the equation for each line and solve for their common point of intersection. After finding the coordinates of the center, we calculate its distance to the vertices to get the radius. Here is what a little bit of courage in the face of unpleasant computation produces for the circumcircle:

$$x = \frac{b_2 \left(a_1^2 + a_2^2\right) - a_2 \left(b_1^2 + b_2^2\right)}{2 \left(a_1 b_2 - a_2 b_1\right)},$$

$$y = \frac{a_1 \left(b_1^2 + b_2^2\right) - b_1 \left(a_1^2 + a_2^2\right)}{2 \left(a_1 b_2 - a_2 b_1\right)},$$

and radius: $\dfrac{\sqrt{\left(a_1^2 + a_2^2\right) \left(\left(a_1 - b_1\right)^2 + \left(a_2 - b_2\right)^2\right) \left(b_1^2 + b_2^2\right)}}{2 \left|a_1 b_2 - a_2 b_1\right|}.$

Orthocenter (x, y)

The computation required for the lines corresponding to the altitudes is less arduous than what we had to do for the circumcenter. We have the slope of the altitudes, which are perpendicular to the edges, from our exercise with the circumcenter. The points on each line are simply the vertices. When we emerge victorious, we have:

$$x = \frac{\left(b_2 - a_2\right) \left(a_1 b_1 + a_2 b_2\right)}{a_1 b_2 - a_2 b_1},$$

$$y = \frac{\left(a_1 - b_1\right) \left(a_1 b_1 + a_2 b_2\right)}{a_1 b_2 - a_2 b_1}.$$

Center, (x, y), and Radius of the Nine-Point Circle

Finally, we want to compute the center (x, y) of the nine-point circle. We know that the circle must pass through the midpoint of each edge. The midpoints are well known to us by now. That is all we need to produce three equations in three unknowns x, y, and the radius r:

$$\left(x - \frac{a_1}{2}\right)^2 + \left(y - \frac{a_2}{2}\right)^2 = r^2,$$

$$\left(x - \frac{b_1}{2}\right)^2 + \left(y - \frac{b_2}{2}\right)^2 = r^2,$$

and $\left(x - \dfrac{a_1 + b_1}{2}\right)^2 + \left(y - \dfrac{a_2 + b_2}{2}\right)^2 = r^2.$

Turning the crank, we get:

$$x = \frac{a_1^2 b_2 + 2a_1 b_1 b_2 + a_2 b_2^2 - 2a_1 a_2 b_1 - a_2 b_1^2 - a_2^2 b_2}{4\left(a_1 b_2 - a_2 b_1\right)},$$

$$y = \frac{a_1^2 b_1 + 2a_1 a_2 b_2 + a_1 b_2^2 - 2a_2 b_1 b_2 - a_1 b_1^2 - a_2^2 b_1}{4\left(a_1 b_2 - a_2 b_1\right)},$$

and radius:
$$\frac{\sqrt{\left(a_1^2 + a_2^2\right)\left(\left(a_1 - b_1\right)^2 + \left(a_2 - b_2\right)^2\right)\left(b_1^2 + b_2^2\right)}}{4\left|a_1 b_2 - a_2 b_1\right|}.$$

We see immediately that the radius of the circumcircle is twice the radius of the nine-point circle. If we then compute the slope of the line containing the circumcenter and the centroid we get

$$\frac{2a_2 b_2\left(b_1 - a_1\right) + a_1 b_2^2 - b_1\left(3a_1\left(a_1 - b_1\right) + a_2^2\right)}{2a_1 b_1\left(a_2 - b_2\right) + a_1^2 b_2 + a_2\left(3b_2\left(a_2 - b_2\right) - b_1^2\right)}.$$

The same value is produced when we calculate the slope of the line containing the orthocenter and the centroid and the line containing the center of the nine-point circle and the centroid. We conclude that the four points are collinear.

We recognize that the expression

$$a_1 b_2 - a_2 b_1$$

appears in the denominators of most of the formulas that we have derived in this section. Since the slopes of the lines containing the edges BC and AC are not the same, we have:

$$\frac{a_2}{a_1} \neq \frac{b_2}{b_1}.$$

This means that

$$a_1 b_2 - a_2 b_1 \neq 0.$$

So we can again set aside zero-in-the-denominator concerns.

We claimed the following earlier: The centers always appear in this order on the Euler line: orthocenter, nine-point center, centroid, circumcenter. The nine-point center is the midpoint of the line segment that connects the orthocenter and the circumcenter. The centroid is always two-thirds of the way from the orthocenter to the circumcenter. These claims can be shown to be true by using the same hard-nosed approach that we just employed to locate the centers. It's all basic algebra, geometry, and grit.

Chapter 15

The Dissertation

When I returned from Germany in 1976, I took some courses to get myself up to speed for graduate study in mathematics. It had been six years since I graduated from Yale College. The first three years, I worked at Gulf Oil as an economic analyst. My major accomplishment at Gulf was a report that gave a forecast of residual fuel oil consumption in the northeastern United States. That was followed by three years in Düsseldorf, most of that time teaching mathematics at the Kaufmännische Schule IV. It was the time in Germany that brought me back to mathematics and graduate school at the University of Pittsburgh.

My regimen in graduate school was typical. I lived in a small basement apartment. A teaching fellowship was sufficient to support my lifestyle. After passing the preliminary examinations, I eventually found myself doing research in a field called *topology*, the study of abstract spaces. There were several topologists at Pitt, including Robert Heath and Frank Slaughter. William Fleissner was the newest addition to the faculty in that discipline and I ended up being his first doctoral student. We were the same age, but Bill had skipped wrestling with residual fuel oil and the Düsseldorf public school system. He allowed me to follow my own interests, which diverged considerably from his. My dissertation grew out of a question that I posed after the two of us had worked through the book by Nagata on dimension theory.[1]

There is the possibility that I will end up walking the reader over the edge of a cliff in this chapter and I feel a certain queasiness. But I am going to move ahead anyway. Sometimes you just take chances. I will attempt to motivate the question that led to my dissertation by talking first about points on the number line: For each point p on the number line and each positive real number r there are precisely two points at the distance r from p. Those points are $p + r$ and $p - r$. We all know this.

Here's the grand question. Is it possible to construct a space S that satisfies the condition that for each point p in S and each non-negative number real number r, there is a unique point in S that is the distance r from p? If so, we'll say that S has the unique-point property.

---

[1] *Modern Dimension Theory*, Jun-iti Nagata, Noordhoff, Groningen, 1965.

To tighten this up just a bit, we will call the device used to measure the distance between two points a *metric*. For a metric d we write $d(a, b)$ to mean "the distance from a to b." Our metric d must satisfy:

$d(a, b) \geq 0$ for all points a and b in the space,

$d(a, b) = 0$ if and only if $a = b$,

$d(a, b) = d(b, a)$ for all points a and b in the space,

$d(a, b) + d(b, c) \geq d(a, c)$ for all points a, b, and c in the space.

The first requirement actually follows from the three listed after it, but it is useful to put it in plain view. The last requirement is called the *triangle inequality* and is roughly equivalent to saying "the shortest distance between two points is a straight line" in plane geometry. (To see that the first requirement follows from the other three, begin by replacing c with a in the triangle equality requirement above.)

Returning to *the question*, I wanted to know if a space S exists that has the unique-point property. This is impossible to visualize, at least it is for me. It was a bit like hunting for Bigfoot for a while, but when I was finally able to determine the existence of such a space S, I showed that there are oodles of them. In fact, there are uncountably many of them in the Cartesian plane, with uncountably many different spatial characteristics. They are not twins or whatever the equivalent term is for infinitely many siblings all split from the same egg. However, I noticed that they did share some "DNA" (three spatial properties) that were intriguing. I will not list them here. I will simply say that Bill Fleissner and I had also worked through Kuratowski's book[2] on topology and I remembered that these properties were also present in a classical space, the space of irrationals. These spaces were just one property short of being the space of irrationals. If I could show that the space of irrationals was in some incarnation a unique-point space, that would imply a precise characterization of the space of irrationals as a unique-point space that has the fourth property, which happens to be called "completeness." Add completeness to the unique-point property and one goes from an uncountable infinity of examples to just one.

The trick was to show that the space of irrationals could be equipped with a unique-point metric that is consistent with its spatial structure. This turned out to be an incredibly elusive trick. Months into my effort, the topologist F. Burton Jones came to Pitt to give a series of lectures. At some point during his visit, my advisor mentioned to him what I was working on. Professor Jones advised that I was likely spinning my wheels. My advisor told me that Jones said that it seemed to him that unique-point spaces are inherently pathological and the space of irrationals is not. A few months

---

[2] *Topology*, Volume I, K. Kuratowski, Academic Press, New York, 1966.

after that, I told Bill that I was going to take a break from mathematics and try debauchery for a couple of months.

Sadly, I did not pass the debauchery preliminary exams and returned to mathematics with a determination to finish my dissertation. I decided to compromise and seek a weaker result; anything to get the degree at this point. I put together some mathematical machinery that I thought would get it done. It involved an algebraic structure that, among other things, has the property that $x + x$ is always equal to 0. I never sufficiently thanked my fellow graduate student Joe Niederberger for the acute observation he made about the initial rudimentary table that I had constructed. Joe had a leaning toward computer science and observed that casting the entries in base 2 lays everything bare. I saw the rhyme and he found the reason. There's no telling how long I might have wandered in the desert otherwise. I later learned that I was using what is called *nim addition*.

But this new machinery did not produce the desired result. After several weeks there was a bolt out of the blue and I realized that the new machinery was precisely what I needed to prove the result I had given up on, namely that the space of irrationals supports a unique-point metric. It took weeks more to write this down completely. There was a dreadful period of four days after that I discovered a hole in my argument. I sat in gloom the whole time with that hollow feeling, which is not uncommon in the mathematical struggle, wondering if I was back to square one. This was especially difficult emotionally because word had already spread through the department that I had broken through. After four days (and nights) I saw my way to patching the problem. The "patch" takes up a significant portion of the paper that I eventually published on this set of results.[3] For months after nailing it down, I would wake up in the middle of the night and spend an hour writing out the main proof, my handwriting getting wild at the end of the ordeal like the printout of a liar's polygraph.

Teoder Przymusinski returned to the University of Pittsburgh from Poland when I was writing the final draft of my dissertation. The course he taught years before, when I was taking prep courses to ready myself for graduate school, was my introduction to topology. The elegance with which he presented the material had impressed me. Teodor read my dissertation and complained that I had buried the main result, the characterization of the space of irrationals, in the back of the paper. I said matter-of-factly that it was a chronological description of the journey.

---

[3]A Metric Characterization of the Irrationals Using a Group Operation, Melvin R. Currie, *Topology and its Applications* **21**, 1985, pp. 223-236, North-Holland.

Chapter 16

The Next Prime Number Is? (Gandhi's Formula)

Here is a gadget that a colleague introduced to me. It is a formula for the nth prime number. I did not believe that such a formula could exist and I had made my belief on this matter clear more than once:

$$p_n = \left\lfloor 1 - \log_2 \left(-\frac{1}{2} + \sum_{d|Q_{n-1}} \frac{\mu(d)}{2^d - 1} \right) \right\rfloor.$$

This formula looks like the sort of contraption that might have been built by Rube Goldberg. J. M. Gandhi pushed it out to the world in 1971.[1] When I first encountered it, I was mystified. Of course, it doesn't take much for this to happen to me. I could not find a proof in the literature that I could stomach. I showed it to a smart mathematician I happened to be working with at the time and his only remark was, "Hmmm...put it in and then take it out, over and over." This terse comment pointed me to a proof. What I am really going to do is begin an explanation and then stop as soon as I think the reader is likely to be saying, "Oh, yeah."

After establishing that the formula is correct, I will allow the reader to reflect on whether or not it is profound. The central mechanism is the garden-variety object, geometric series, which we discussed in Chapter 5. Before we get down to brass tacks, we need to make sure that the notation is understood. We'll start with the outermost piece and work our way in.

$\lfloor x \rfloor$ is the greatest integer less than or equal to x

(also called the *floor function*).

In order to find the nth prime, the formula requires that we know all of the primes less than the nth prime and uses them to compute

Q_{n-1} : the product of the first $n-1$ prime numbers. Q_0 is defined to be 1.

---

[1] J. M. Gandhi, Formulae for the Nth prime. In "Proceedings of the Washington State University Conference on Number Theory," J. H. Jordan and W. A. Webb editors, Department of Mathematics, Washington State University, Pullman, Washington, 1971. pp. 96-106

The reader should note that in our summation we are summing over all the divisors d of Q_{n-1}. That is indicated by the notation $d|Q_{n-1}$.

Finally, we have to define the Möbius function, which is commonly denoted by μ.

An integer n is said to be square-free if it is not divisible

by the square of a prime.

$\mu(n) = 0$, if n is not square-free.

$\mu(n) = -1$, if n is square-free and the product of an odd number of primes.

$\mu(n) = 1$, if n is square-free and the product of an even number of primes.

In particular, $\mu(1) = 1$, since 1 is the product of zero primes. Zero is even.

All of the divisors of Q_{n-1} are necessarily square free, so $\mu(d)$ will only take on the values 1 or -1 in Gandhi's formula, depending on whether or not the divisor d has an even or an odd number of prime factors.

It might be useful to confirm that the formula tells us that the first prime is 2. Plugging things in we get

$$
\begin{aligned}
p_1 &= \left\lfloor 1 - \log_2\left(-\frac{1}{2} + \sum_{d|Q_0} \frac{\mu(d)}{2^d - 1}\right)\right\rfloor \\
&= \left\lfloor 1 - \log_2\left(-\frac{1}{2} + 1\right)\right\rfloor \\
&= \left\lfloor 1 - \log_2\left(\frac{1}{2}\right)\right\rfloor \\
&= 1 - (-1) \\
&= 2.
\end{aligned}
$$

We begin "unpacking" Gandhi's formula. (This is a term used by social scientists these days, when discussing language usage. I thought I might try it out in this context, since Gandhi's formula appears to be packed with meaning.)

$$
p_n = \left\lfloor 1 - \log_2\left(-\frac{1}{2} + \sum_{d|Q_{n-1}} \frac{\mu(d)}{2^d - 1}\right)\right\rfloor.
$$

We claim that the first term in the summation is 1. That's because 1 is always a divisor and $\mu(1) = 1$. Adding 1 and $-1/2$, we get the slightly transformed formula

$$
p_n = \left\lfloor 1 - \log_2\left(\frac{1}{2} + \sum_{\substack{d|Q_{n-1} \\ d>1}} \frac{\mu(d)}{2^d - 1}\right)\right\rfloor.
$$

Remembering our observations about geometric series, we realize that $\frac{1}{2}$ can be replaced by

$$\frac{1}{2^2} + \frac{1}{2^3} + \frac{1}{2^4} + \frac{1}{2^5} + \frac{1}{2^6} + \cdots.$$

This makes the formula messier, but it will quickly make it more transparent. We have

$$p_n = \left\lfloor 1 - \log_2 \left(\frac{1}{2^2} + \frac{1}{2^3} + \frac{1}{2^4} + \frac{1}{2^5} + \frac{1}{2^6} + \cdots + \sum_{\substack{d|Q_{n-1} \\ d>1}} \frac{\mu(d)}{2^d - 1} \right) \right\rfloor.$$

A little more fiddling will make things even clearer. Since

$$\frac{1}{2^d - 1} = \frac{\frac{1}{2^d}}{1 - \frac{1}{2^d}} = \frac{1}{2^d} + \frac{1}{2^{2d}} + \frac{1}{2^{3d}} + \frac{1}{2^{4d}} + \frac{1}{2^{5d}} + \cdots,$$

we can transform our expression further to get

$$p_n = \left\lfloor 1 - \log_2 \left(\frac{1}{2^2} + \frac{1}{2^3} + \frac{1}{2^4} + \frac{1}{2^5} + \frac{1}{2^6} + \cdots \right.\right.$$

$$\left.\left. + \sum_{\substack{d|Q_{n-1} \\ d>1}} \mu(d) \left(\frac{1}{2^d} + \frac{1}{2^{2d}} + \frac{1}{2^{3d}} + \frac{1}{2^{4d}} + \frac{1}{2^{5d}} + \cdots \right) \right) \right\rfloor.$$

Let's calculate p_2 using our unpacked version of Gandhi's formula. Q_1 is simply $p_1 = 2$ and $\mu(2) = -1$:

$$p_2 = \left\lfloor 1 - \log_2 \left(\frac{1}{2^2} + \frac{1}{2^3} + \frac{1}{2^4} + \frac{1}{2^5} + \frac{1}{2^6} + \cdots \right.\right.$$

$$\left.\left. + \sum_{\substack{d|2 \\ d>1}} \mu(d) \left(\frac{1}{2^d} + \frac{1}{2^{2d}} + \frac{1}{2^{3d}} + \frac{1}{2^{4d}} + \frac{1}{2^{5d}} + \cdots \right) \right) \right\rfloor.$$

Following the instructions, we get

$$p_2 = \left\lfloor 1 - \log_2 \left(\frac{1}{2^2} + \frac{1}{2^3} + \frac{1}{2^4} + \frac{1}{2^5} + \frac{1}{2^6} + \cdots \right.\right.$$

$$\left.\left. - \left(\frac{1}{2^2} + \frac{1}{2^4} + \frac{1}{2^6} + \frac{1}{2^8} + \frac{1}{2^{10}} + \cdots \right) \right) \right\rfloor$$

$$= \left\lfloor 1 - \log_2 \left(\frac{1}{2^3} + \frac{1}{2^5} + \frac{1}{2^7} + \frac{1}{2^9} + \frac{1}{2^{11}} + \cdots \right) \right\rfloor = \left\lfloor 1 - \log_2 \left(\frac{1}{6} \right) \right\rfloor = 3.$$

The next-to-last equality follows from what we know about the sum of a geometric series. Recognizing that $\frac{1}{6}$ lies between 2^{-2} and 2^{-3} completes the calculation.

The mystery is fading. Gandhi's formula just puts into a neat package an approach that has been known since antiquity called the Sieve of Eratosthenes. What we are doing is subtracting all the terms with exponents that are multiples of primes in the product Q_{n-1} (product of primes less than p_n). What we have left is

$$\frac{1}{2^{p_n}} + \text{ some distinct powers } (> p_n) \text{ of } \frac{1}{2}.$$

In our computation of p_2, the distinct powers were the odd integers greater than 3.

Returning to the general case, p_n, each of these distinct powers of $\frac{1}{2}$ is less than $\frac{1}{2^{p_n}}$. Recalling what we know about geometric series, their sum must be less than $\frac{1}{2^{p_n}}$.

If we take the logarithm base 2 of such a number, multiply it by minus one, add one, and take its floor, we get p_n. Okay, that was too fast. Let's slow it down and write it out. We will denote by ε "the sum of the distinct powers of $\frac{1}{2}$."

$$\frac{1}{2^{p_n}} < \frac{1}{2^{p_n}} + \varepsilon < \frac{1}{2^{p_n}} + \frac{1}{2^{p_n}} = \frac{1}{2^{p_n-1}}.$$

Taking logarithms in base 2 we get

$$-p_n < \log_2(\frac{1}{2^{p_n}} + \varepsilon) < 1 - p_n,$$

$$p_n > -\log_2(\frac{1}{2^{p_n}} + \varepsilon) > p_n - 1,$$

$$1 + p_n > 1 - \log_2(\frac{1}{2^{p_n}} + \varepsilon) > p_n.$$

Finally, taking the floor of the middle term, we get p_n.

The fly in the ointment is that if we do this for each divisor of Q_{n-1}, we end up subtracting out too much, when, as opposed to the example above, we have more than one prime preceding the prime we are trying to calculate. All the common multiples of prime divisors of Q_{n-1}, p_j, and p_k, will be subtracted twice. We have to put them back once! That's where divisors, d, of the form $d = p_j p_k$ come into play. We know that $\mu(p_j p_k) = 1$. As it turns out, by adding in those powers of $\frac{1}{2}$ we may produce another error. So we fix this by subtracting out the multiples of the triples, p_h, p_k, p_j, that appear too often. Remember that $\mu(p_h p_j p_k) = -1$.

Do we ever balance the books? We do. We recognize that all that is really in play here is the Inclusion-Exclusion Principle: "put it in, take it out...."

If we have two sets, A and B, how many elements do we have in their union? We can reason to the answer quickly. It's just the number of elements in A plus the number of elements in B minus the number of elements that they have in common. (In preparation for what follows, some readers might find it useful to consult Appendix A if they are unfamiliar with the notation and terms associated with set operations)

Let $|S|$ stand for the number of elements in S.

Then

$|A \cup B| = |A| + |B| - |A \cap B|$, where stands $A \cap B$ for A "and" B.

Let the set A correspond to the multiples of the first prime and B to the multiples of the second prime.

If our Q_{n-1} is just the product of the first two primes, we will have made all the corrections that we need to make after we add back in the error we made by subtracting their common multiples twice. We push it one more step to the case of the first three primes, so that we see the rhythm of Inclusion-Exclusion.

We want the formula for

$$|A \cup B \cup C|.$$

Let

$$M = A \cup B.$$

Using what we know about the union of two sets, we can write

$$|M \cup C| = |M| + |C| - |M \cap C| = |A| + |B| - |A \cap B| + |C| - |(A \cap C) \cup (B \cap C)|$$
$$= |A| + |B| + |C| - |A \cap B| - |A \cap C| - |B \cap C| + |A \cap B \cap C|.$$

We could use the same machine to prove by induction that the Inclusion-Exclusion Principle holds for any finite number of sets. We have already done it for two sets and three sets, so we will omit the formality.

Let

$$G = \left\{ \frac{1}{2^2}, \frac{1}{2^3}, \frac{1}{2^4}, \frac{1}{2^5}, \frac{1}{2^6}, \frac{1}{2^7}, \cdots \right\}.$$

Assume that we want to remove each element of G that has an exponent that is a multiple of one of the first three primes p_1, p_2, p_3, which are 2, 3, and 5, respectively. Let

$$A = \left\{ \frac{1}{2^{p_1}}, \frac{1}{2^{2p_1}}, \frac{1}{2^{3p_1}}, \frac{1}{2^{4p_1}}, \frac{1}{2^{5p_1}}, \frac{1}{2^{6p_1}}, \cdots \right\},$$

$$B = \left\{ \frac{1}{2^{p_2}}, \frac{1}{2^{2p_2}}, \frac{1}{2^{3p_2}}, \frac{1}{2^{4p_2}}, \frac{1}{2^{5p_2}}, \frac{1}{2^{6p_2}}, \cdots \right\},$$

$$C = \left\{ \frac{1}{2^{p_3}}, \frac{1}{2^{2p_3}}, \frac{1}{2^{3p_3}}, \frac{1}{2^{4p_3}}, \frac{1}{2^{5p_3}}, \frac{1}{2^{6p_3}}, \cdots \right\}.$$

To make life simpler, let's agree to use the plus sign in (*) below to indicate "put it in" and the minus sign to indicate "take it out." Then our Inclusion-Exclusion Principle for three sets tells us that we can carry out this extraction by performing the following operations:

(*) $G - A - B - C + (A \cap B) + (A \cap C) + (B \cap C) - (A \cap B \cap C).$

But this is exactly what's going on (put it in and take it out) when we compute p_4 using the unpacked Gandhi's formula below:

$$p_4 = \left\lfloor 1 - \log_2\left(\frac{1}{2^2} + \frac{1}{2^3} + \frac{1}{2^4} + \frac{1}{2^5} + \frac{1}{2^6} + \cdots \right.\right.$$

$$\left.\left. + \sum_{\substack{d|2\cdot3\cdot5 \\ d>1}} \mu(d)\left(\frac{1}{2^d} + \frac{1}{2^{2d}} + \frac{1}{2^{3d}} + \frac{1}{2^{4d}} + \frac{1}{2^{5d}} + \cdots \right) \right) \right\rfloor.$$

I'm ready to say, "Oh, yeah." How about you?

Chapter 17

Bulgarian Solitaire

I first met Bob Cordwell when he was in the seventh grade. I was Chief of the Cryptographic Research and Design Division (The Codemakers) at the National Security Agency. His father, Bill Cordwell, was a physicist on assignment to NSA from Sandia National Labs and working in my division. Bob was precocious. He was taking a course in differential equations at the time. His senior year in high school, 2005, Bob was on the six-member team that represented the United States in the International Mathematical Olympiad, where he won a gold medal. Bob was a top finisher in the Intel Science Talent Search that year as well. The Currie and the Cordwell families stayed in touch over the years. When Bob was in his mid-twenties I came to him with a question. I had just been introduced to Bulgarian Solitaire. To my great disappointment, I was not able to prove the central claim about this intriguing system. As is usually the case, I was not happy with the proofs available in the literature. I described Bulgarian Solitaire to Bob as well as my level of frustration. A few weeks later he sent me a proof[1] from The Book.

For the uninitiated, let's begin with a description of Bulgarian Solitaire. The phenomenon burst onto the mathematics scene in the early eighties, but the history already has the feel of folklore. We can imagine ten children's blocks sitting on a table arranged in an arbitrary number of stacks. Let's say that we have a stack of 6 and a stack of 4. The transformation of interest simply requires that we remove one block from each of our stacks to form a new stack (stacks with zero blocks are ignored). Applying the transformation to stacks 6 and 4 results in stacks of 5, 3, and the newly formed stack of 2. One more iteration yields 4, 2, 1, and 3. Since we are not at the moment concerned with ordering the stack sizes, it is clear that equilibrium has been reached. Applying the transformation again will not alter the stack sizes, 1, 2, 3, and 4. Jørgen Brandt wrote the earliest paper on this topic of which I am aware. The passage to equilibrium carries his name. Brandt's Equilibrium Theorem (BET) states that for any triangular number $t_k = \frac{k(k+1)}{2}$ and any initial configuration of stacks that sum to t_k, repeated

---

[1] A few years after Robert Cordwell discovered this proof, Andrew Frohmader communicated a proof to me that was essentially identical.

applications of the transformation will eventually result in the equilibrium arrangement of stack sizes $1, 2, 3, \ldots, k$.

There are already proofs of BET available to the reader and there would not be any reason to put this one forward, if it did not offer something different, at least on an aesthetic level. Our aim is not only to show that BET holds, we would also like to provide some intuition for *why* convergence to a specific equilibrium is inevitable. We will judge our effort a success if the reader is indeed left with an intuitive feel and an appreciation for the truth of BET.

Cordwell's Approach

Anyone who thinks about this problem will quickly see that, despite the simplicity of the transformation, attempting to capture its description using standard mechanisms is a challenge. We have chosen a framework that provides a way to track the changes in the status of our configuration as we iterate. The framework will be the first quadrant of the xy-plane. In Figure 1, the initial configuration that we discussed in the introduction for t_4 is displayed.

The number in the interior of each block is simply the sum $x + y - 1$, where (x, y) represents the coordinates of the upper right vertex of the block, which we will call the *pertinent vertex*. We'll call $x + y - 1$ the *weight* of the block. The weight of the configuration (the sum of the weights of the blocks in the configuration) in Figure 1 is 35. For the sake of the argument that follows, we implement our transformation as follows: All blocks that do not lie on the x-axis shift one square to the right and down one. All blocks that lie on the x-axis are re-positioned so that their upper right corner is $(1, x)$ instead of $(x, 1)$. This can be visualized as rotating the bottom row of blocks clockwise through an arc of 270°, so that instead of taking one block from the top of each stack to form the new stack, we have taken one block from the bottom of each one. Note that this is completely within the rules of the game. We could also view our transformation as performing circular shifts on the *diagonals* that only contain blocks of a given weight. This latter view will be helpful later, when we do some modular arithmetic. (A review of modular arithmetic is provided in Appendix A.) See Figure 2 for the result of the first iteration.

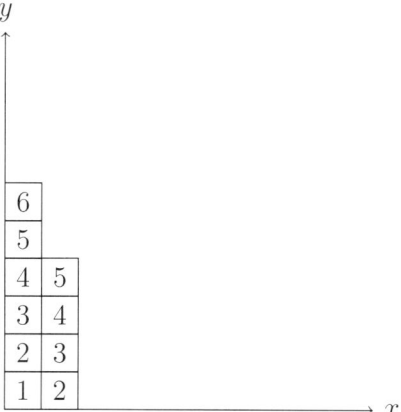

FIGURE 1.

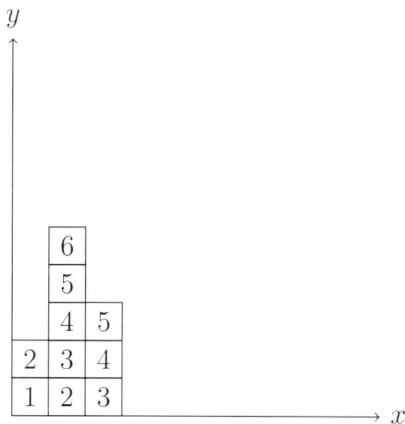

FIGURE 2.

We recognize that the transformation is weight-preserving by construction, since in the case of the set of blocks that are not on the x-axis, the vertex of interest is moved from (x, y) to $(x+1, y-1)$. The other case, $(x, 1)$ to $(1, x)$, obviously doesn't change the weight either.

We perform the transformation again to get the configuration in Figure 3.

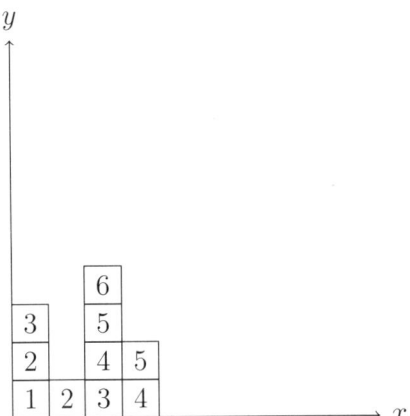

FIGURE 3.

We have now arrived at the state that, without the framework, would be the end of the game. However, there turns out to be value in pressing on. One more iteration produces Figure 4.

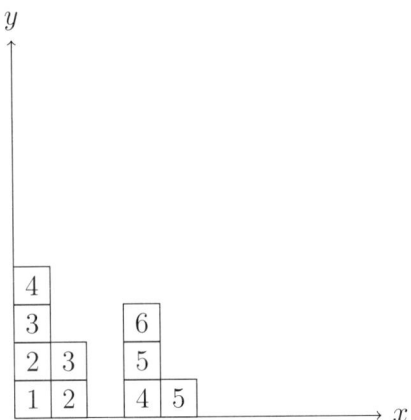

FIGURE 4.

Well, it isn't tidy. Our block configuration is not connected. We remedy this by sliding the four rightmost blocks one unit to the left. Of course, all four of them lose weight as a result, and we have Figure 5.

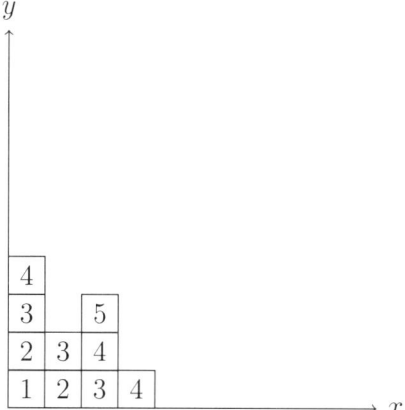

FIGURE 5.

The weight has fallen from 35 to 31, as a result of the leftward slide. If you like, you can view the slide as being the result of attraction. Nature abhors a gap. (In the case of more than one gap at a time, we do what one might expect. We slide leftward (rigidly) until our configuration is connected.)

To spare ourselves the tedium, we simply note that two more iterations take us to Figure 6.

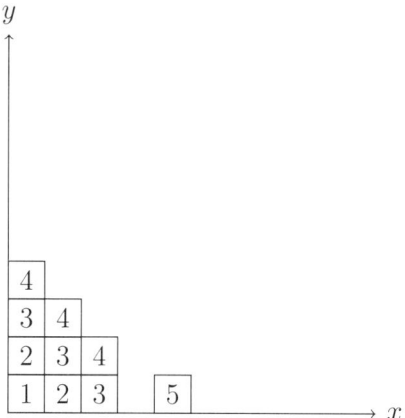

FIGURE 6.

A slide produces Figure 7.

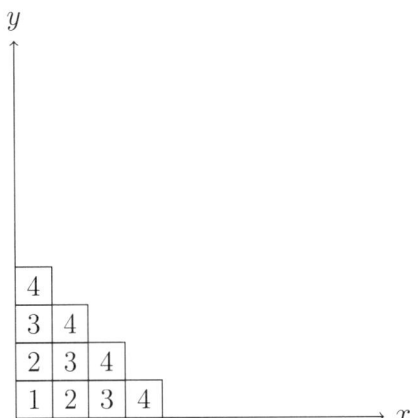

It's clear that no further progress can be made. We have reached a weight of 30. It is also obvious that the configuration in Figure 7 is the only arrangement of ten blocks that can have this weight, which is also the minimum weight for ten blocks. We'll call the configuration in Figure 7 the *minimum weight configuration* or **mwc**. At this point we note that our implementation of the transformation carries the blocks through circular cyclic shifts. Blocks of the same weight stay on a diagonal.

The general proof of BET for any triangular number of blocks is effectively embedded in this example. The minimum weight for any triangular number t_k of blocks is the sum of the squares of the first k integers, which is computed using an identity that we proved in Chapter 5:

$$\sum_{n=1}^{k} n^2 = \frac{k(k+1)(2k+1)}{6}.$$

We only need to show that weight loss through leftward attraction will eventually occur when the transformation is applied for any initial configuration other than the **mwc**, which covers precisely all squares corresponding to a weight less than or equal to k. Successive configurations can never increase in weight. We saw in our example that repeated application of the transformation forced the blocks to *fall* into the **mwc**. A leftward slide occurs when there is a block on the x-axis, with pertinent vertex at $(x, 1)$ $(x > 2)$, but there is not a block with pertinent vertex at $(x - 1, 1)$. We now argue that a leftward slide must always occur eventually if the configuration is not the **mwc**.

If a configuration for t_k is not the **mwc**, then at least one block must have weight greater than k, so at least one diagonal must be incomplete. Suppose we have a configuration with a block of weight m $(m > 2)$ and the $(m - 1)$-weight diagonal is the lowest-weight incomplete diagonal. Then for some (s, t) with

$$s + t - 1 = m - 1,$$

there is no block whose pertinent vertex is at (s, t). The block of weight m and the gap on the weight-$(m - 1)$ diagonal are on track to produce a gap on the x-axis after a certain number of steps, c. Let the block of weight m have pertinent vertex at (u, v). What are we looking for? Our circular-shift view of the transformation is this. We want to find c such that

$$c + s = 0 \quad \text{mod } m - 1,$$

$$c + u = 0 \quad \text{mod } m.$$

(There is a review of modular arithmetic in Appendix A.) The fact that the integers m and $m - 1$ are consecutive guarantees the existence of a solution. The solution tells us how many steps we must take before we will have a slide involving this block of weight m. That is assuming that a slide has not already occurred for some other pairing of block and gap in the course of repeated application of the transformation.

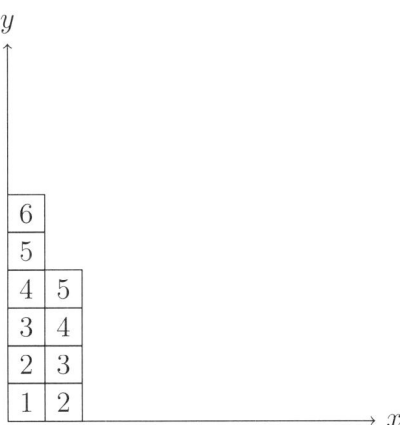

FIGURE 8.

Going back to the initial setting in our example in Figure 8, we see that there are several incomplete diagonals. If we, for instance, want to compute how quickly a gap will open to the left of a weight-4 block, we solve the following pairs of modular equalities, noting that the pertinent vertex of the empty space on the weight-3 diagonal is $(3, 1)$ and the two occupied weight-4 positions have pertinent vertices $(1, 4)$ and $(2, 3)$. To compute when a weight-4 block will reach the x-axis at the same time as the empty space on the weight-3 diagonal, we find c that satisfies one of the following pairings of modular equations:

$$c + 3 = 0 \quad \text{mod } 3 \qquad\qquad c + 3 = 0 \quad \text{mod } 3$$
$$\text{and} \qquad\qquad\qquad\qquad \text{and}$$
$$c + 2 = 0 \quad \text{mod } 4 \qquad\qquad c + 1 = 0 \quad \text{mod } 4.$$

This is equivalent to

$$c = 0 \quad \mod 3 \qquad\qquad c = 0 \quad \mod 3$$
$$\text{and} \qquad\qquad\qquad \text{and}$$
$$c = 2 \quad \mod 4 \qquad\qquad c = 3 \quad \mod 4.$$

We see that the smallest non-negative solutions are 6 and 3, respectively. We conclude that there are three iterations to the first gap, assuming that a gap doesn't occur earlier next to a weight-5 or weight-6 block. Since we have already worked through this example from beginning to end, we needn't burden ourselves in this case with solving the other pairs of modular equalities for weights greater than four.

Summing it up, we are guaranteed a gap unless the configuration is the **mwc**, in which case no gap can occur. All roads ultimately lead to the **mwc**.

The Non-Triangular Case

What about the non-triangular case? Suppose that we had 12 blocks. We claim that it's easy to see using the same argument as above that the distribution of blocks will eventually fall into a configuration that minimizes the weight. In the case of 12 blocks, the minimum weight is 40. So, although we don't get a fixed configuration at the end of the game, we do converge to a cycle of configurations, all of which have the same weight.

Suppose that

$$m = t_k + j, \text{ where}$$
$$0 < j < k + 1.$$

Then repeated application of the transformation will force the blocks into a cycle of configurations, all of weight

$$\frac{k(k+1)(2k+1)}{6} + j(k+1).$$

We can visualize this as the t_k **mwc** with the addition of j weight-$(k+1)$ blocks cycling along the weight-$(k+1)$ diagonal. The number of distinct cycles corresponds to the number of necklaces that can be formed using j white beads and $k+1-j$ black beads. So, if j equals 1 or k, only one cycle is possible. For $j = 0$ or $j = k+1$, we have the kth and $(k+1)^{\text{th}}$ triangular numbers.

Let's look at an eleven-block case. What we are going to see after a few applications of the transformation, and a couple of leftward slides, is the **mwc** for triangular number 10 with one lonely weight-5 block doomed to cycle on the weight-5 diagonal by itself for eternity.

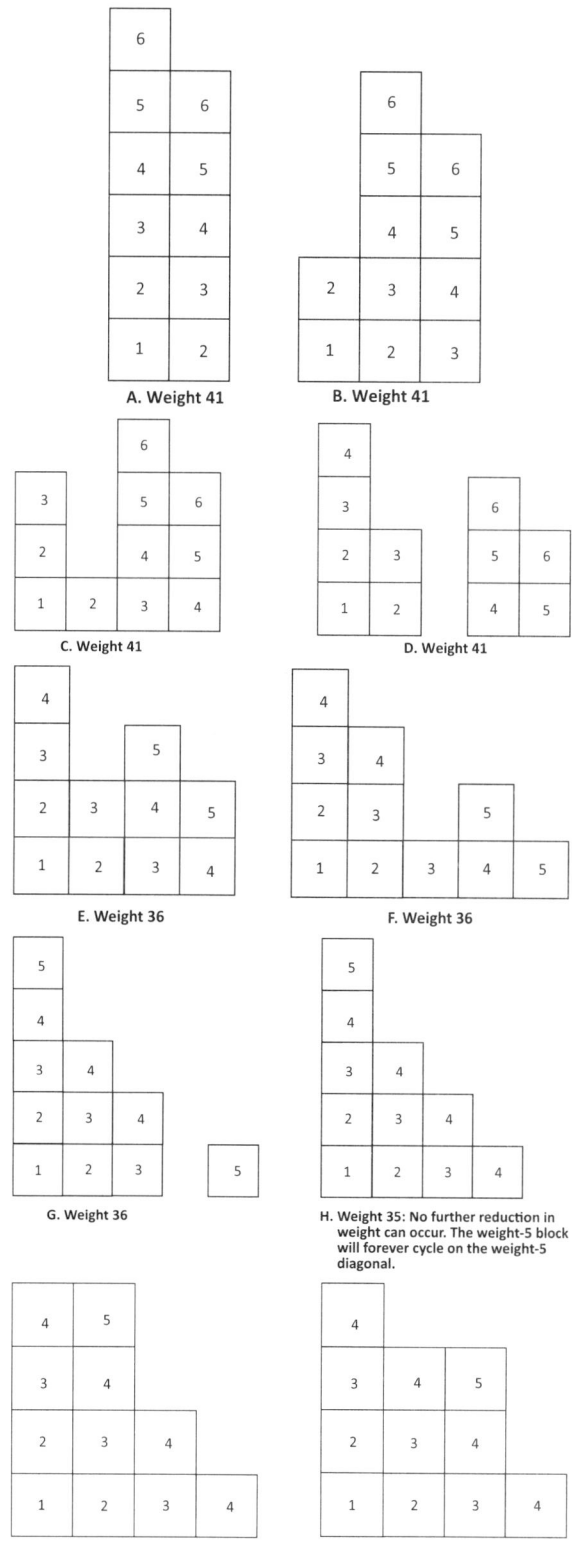

A. Weight 41

B. Weight 41

C. Weight 41

D. Weight 41

E. Weight 36

F. Weight 36

G. Weight 36

H. Weight 35: No further reduction in weight can occur. The weight-5 block will forever cycle on the weight-5 diagonal.

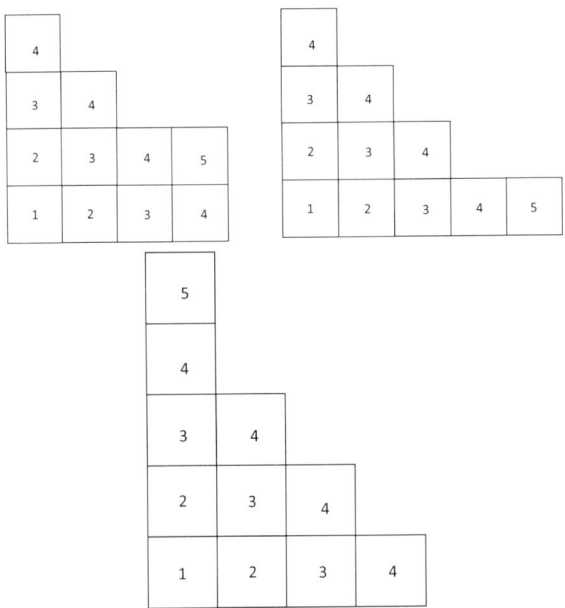

The Gateway and the Maximal Number of Iterations to the mwc

In this section we'll sketch an explanation for why the maximal number of iterations to the **mwc** is $k^2 - 1$.

For the triangular number t_k, the only possible preceding configurations for the stack sequence $1, 2, 3, \ldots, k$ are either itself or $k+1, k-1, k-2, \ldots, 2$. That's because only the stacks of size k and $k-1$ could be the stack formed by having reduced the remaining stacks by one in a preceding configuration. A stack of size $k-2$ or smaller is too small to have come from taking one block from each of at least $k-1$ stacks. If it's the stack of size k, the preceding configuration must have been $1, 2, 3, \ldots, k$. Otherwise, the preceding configuration clearly must have been $k+1, k-1, k-2, \ldots, 2$. At this point we call a timeout and suggest to the reader that a good antidote to the formality here is to keep in mind triangular numbers 10 and 15. Their gateways are $5, 3, 2$ and $6, 4, 3, 2$, respectively. Now you're good to go.

For $k > 2$, call the configuration $k+1, k-1, k-2, \ldots, 2$ the *gateway*. Once successive iterations have reached the gateway, we claim that there are precisely k additional iterations to the **mwc**. We can see this by induction and use t_3 as the basis. The gateway is $4, 2$:

$$4, 2 \to 2, 3, 1 \to 3, 1, 2 \to 3, 2, 1.$$

We required three steps to the **mwc**, so the claim holds for $k = 3$.

We now assume it's true for k and consider $k + 1$:

$$k+2, k, k-1, k-2, \ldots, 2 \to k, k+1, k-1, k-2, \ldots, 2, 1$$
$$\to k+1, \underline{k-1, k, k-2, \ldots, 1} \cdots \to k+1, k, k-1, \ldots, 1.$$

We see that after the first iteration the gateway for t_k is embedded in the sequence and the course that it follows, as we iterate, is exactly the same as if it were standing alone as the gateway for t_k. As a result there are k iterations after the second arrow and a total of $k + 1$ iterations after the gateway for t_{k+1}. Again, keep in mind triangular numbers 15 and 10. Their gateways are $6, 4, 3, 2$ and $5, 3, 2$, respectively. This is a see-it-and-believe-it section.

Yes, the reader might have noticed that we assumed an ordering on the gateways, and order *does* count in this discussion. However, we suspect that the reader will then recognize that the argument for any other ordering is completely analogous, which relieves the author of any further responsibility.

Kiyoshi Igusa[2] proved that convergence to the Brandt Equilibrium (not observing the order of the numbers 1 through k) occurs after no more than $k^2 - k$ iterations. Since we are taking order into account, the argument above says that the maximal number of iterations to the **mwc** occurs after no more than $k^2 - 1$ iterations. We demonstrated that there are k iterations from the gateway to the **mwc**. But the Brandt Equilibrium is reached immediately after we apply the transformation to the gateway and there are $k-1$ iterations after that:

$$k^2 - k + k - 1 = k^2 - 1.$$

It is now easy to produce a canonical example of a configuration of t_k blocks that requires the maximal number of iterations to the **mwc**. (See Figure 9 below.) It's $k - 1, k - 1, k - 2, k - 3, \ldots, 2, 1, 1$ for $k > 2$. This is a little strange. The configuration only differs from the **mwc** in the placement of one block. And its weight is only one unit more than the **mwc** for a t_k. Nonetheless, it requires $k^2 - 1$ iterations to reach the **mwc**.

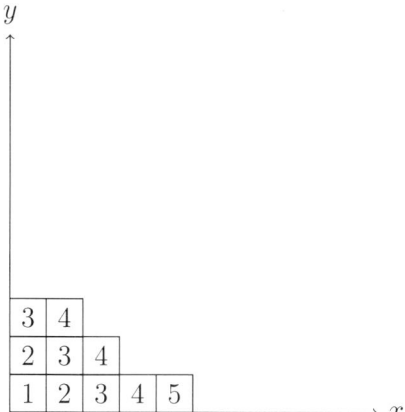

FIGURE 9.

[2]"Solution of the Bulgarian Solitaire Conjecture", Kiyoshi Igusa, *Mathematics Magazine*, 58:5, 2018, 259-271, DOI: 10.1080/0025570X.1985.11977198

In the case of t_4, the number of iterations c corresponds to the missing square in the weight-four position reaching the x-axis at the same time that the weight-five block does. We solve the following system:

$$c = 0 \quad \mod 5 \text{ and}$$
$$c + 1 = 0 \quad \mod 4.$$

This is equivalent to:

$$c = 0 \quad \mod 5 \text{ and}$$
$$c = 3 \quad \mod 4.$$

The smallest nonnegative solution is

$$15 = 4^2 - 1.$$

In general, we would solve:

$$c = 0 \quad \mod k + 1 \text{ and}$$
$$c = k - 1 \quad \mod k.$$

The solution is $c = k^2 - 1$.

The framework that was introduced in this paper has a very physical flavor, which is not apparent in the original straightforward description of Bulgarian Solitaire. Yet, the particular convention for carrying out the transformation and the slide are clearly valuable in understanding the convergence to equilibrium for the triangular number and the structure of the cycles for the non-triangular case. We see no such "strings" when we are carrying out the exercise with blocks on a table. This is reminiscent of the configuration of electrons in shells about the nucleus; a kind of Aufbau Principle, but without apparent physical forces at play.

Chapter 18

Which is Bigger? (a^b versus b^a)

Every now and then, I see problems that have been used in various mathematics contests, and every now and then, I am able to solve one of them, but almost never in the amount of time that would make me competitive. I am just not a mathematical gunslinger. There is one type of problem that I have seen more than once. It looks like this:

Which is bigger, 7.01^7 or $7^{7.01}$?

It was always pretty clear that using a calculator was not allowed. I had not permitted myself to be seduced by such problems until I saw the question in terms of two numbers that I really cared about:

Which is bigger, e^π or π^e?

I went after this question and was able to show in a few different ways that e^π is the larger of the two. What began to gnaw at me is that none of the approaches seemed to be specific to π.

Here is the most useful approach for this discussion; the one that made it clear to me what was going on. Suppose we want to compare a^b with b^a. We will assume that a and b are both positive. The "log" used below is the natural log, logarithm base $e = 2.71828\ldots$. The question marks below indicate that we do not know which expression is bigger:

$$a^b \,?\, b^a,$$

$$b \log a \,?\, a \log b.$$

Hoping that I can discern an inequality that I can walk back to my original comparison, I multiply on both sides of the question mark by $1/ab$:

$$\frac{\log a}{a} \,?\, \frac{\log b}{b}.$$

This prods us to take a look at the function

$$f(x) = \frac{\log x}{x}.$$

We do this because we don't know what else to do. Desperation pays off, because I (and maybe you) have calculus in my tool kit. We would like to know where f achieves its maximal value, if it achieves a maximal value at all. Take the derivative and set it to zero to find out where the tangent line is horizontal!

$$f'(x) = \frac{1 - \log x}{x^2} = 0 \text{ only at } x = e.$$

Since we also manage to notice that the derivative is negative for $x > e$ and positive for $0 < x < e$, our function is maximized for the values of concern to us at $x = e$. Now we can slay the dragon. Our maximum occurs at e. So

$$f(e) = \frac{\log e}{e} = \frac{1}{e} > \frac{\log a}{a}, \text{ for any } a > 0, a \neq e.$$

We backtrack.

$$a > e \log a = \log a^e.$$

Finally,

$$e^a > e^{\log a^e} = a^e, a \geq 0, a \neq e. \text{ (We can obviously include } a = 0.)$$

Choosing π was just a red herring. We can pick any non-negative number not equal to e. This fact characterizes e. But the method of proof tells us more. Since our function f is increasing for $x < e$ and decreasing for $x > e$, we see that

$$0 \leq a < b < e \Rightarrow a^b < b^a,$$
$$e < a < b \Rightarrow a^b > b^a.$$

In general we cannot make a determination immediately, if one of our numbers is less than e and the other is greater.

We can pull one more nugget from this discussion. If a and b are distinct integers and $a^b = b^a$, then the pair a and b must be 2 and 4. We see this in the following way. For any integer $n > 4$

$$\frac{\log 4}{4} > \frac{\log n}{n},$$

because f is decreasing for $x > e = 2.71828\ldots$ It follows that

$$\frac{\log 2}{2} = \frac{\log 4}{4} > \frac{\log n}{n},$$

so $2^n > n^2$ if $n > 4$ and, since $2^3 < 3^2$, we have completed the case for when one of our integers is 2.

Clearly, if we pick two integers greater than e, they cannot satisfy $a^b = b^a$, again because of the decreasing nature of f. That leaves us with 2 and 4 being the unique pair that satisfies $a^b = b^a$.

At any rate, it is absolutely clear now that

$$7^{7.01} > 7.01^7.$$

Laying the groundwork to show that 2 and 4 are the only integers that satisfy $a^b = b^a$ was not what I would call easy. It reminds me of another question. How many ways can a rectangle be constructed so that its area equals its perimeter, if the length of each edge must be a whole number? Here are two examples of rectangles that meet the requirements.

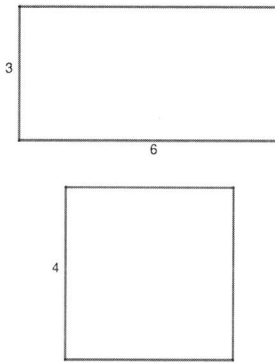

Are there others? Initially one might think that this is as difficult to address as our $a^b = b^a$ inquiry. However, that is not the case. Our rectangles with dimensions $a \times b$ must satisfy

$$ab = 2a + 2b.$$

Following our nose, we solve for a and get

$$a = \frac{2b}{b - 2} = 2 + \frac{4}{b - 2}.$$

Then we remember that a and b must be positive integers. Right off the bat that forces $b - 2$ to be a divisor of 4. After eliminating $b = 1$, 2, and 5, we see that the only positive integers that do the job as choices for b are 3, 4, and 6, corresponding to 6, 4, and 3 for a, respectively. If b is bigger than 6, $b - 2$ is not a divisor of 4.

Cute and not hard. You can never be sure what the nature of the challenge will be when you open your mouth and pose a question.

Chapter 19

Fascinating

Strange things happen in life. One of the stranger happenings for me was when Donald J. Newman appeared in the Cryptographic Research and Design Division at NSA (*The Codemakers*), when I was heading it up. Somehow, he came in through our Sabbatical Program, despite not being on sabbatical. He was in emeritus status at Temple University. Newman began making his reputation early in life. He was a Putnam winner three years in a row, while at CUNY. When I asked him what happened his fourth year he told me that he was not eligible in what would have been his senior year. He finished college in three years and had moved on to graduate work at Harvard.

It was like having my grandfather at work with me. That's how it felt, although Donald was only eighteen years older than I was. I remember the day that he landed in the Division, looking rather tired, but nonetheless dapper in his leather jacket. I had first heard of him in the book about John Nash, *A Beautiful Mind*. Newman was a lover of fast cars. He told me that when he was a young man, he would race airplanes down a runway near where he worked as they took off. He was considered one of the great problem solvers of his era. People would say that if a problem could be solved in twenty-four hours, Newman could do it. The proof that he discovered for the prime number theorem is considered the quintessential one. I once showed him "my" sequence for π. He had never seen it. I also gave him some estimates I had made of the error that would result when using the nth element in the sequence to approximate π. He must not have looked at the stuff for a few days. When he did, he came flying into my office as fast as his legs and his cane could carry him. The way he whipped around the corner, one might have thought that his hair was on fire and he was trying to put it out. The estimates were wrong, he said. "Wrong and crazy!" He probably tacked on a few more adjectives that I can't remember. I don't think I said anything. He flew out of my office and back to his desk. It could not have been more than five minutes later when he charged back in, said that he was mistaken and then left again. Before he dashed away, I might have said that I had been considering the sequence for a long time.

At any rate, I decided to start a seminar on analytic number theory a few weeks after Donald's arrival. I felt that we had to take advantage of the

great man's presence. Our first meeting was on a Monday. Donald had what
I think he believed would be a treat for us. Somehow, he had gotten wind
of a problem with which two mathematicians in another corner of NSA had
been playing around. They had seen empirically that the average value of
the fractional parts of an integer (see below for a definition) does not seem
to converge to $\frac{1}{2}$ as the integer gets large. In fact, the average value gets
close to a value that is approximately 0.422784.

We denote the greatest integer less than or equal to x by $\lfloor x \rfloor$. The average
value of the fractional parts of n is defined as

$$\frac{1}{n} \sum_{k=1}^{n} \left(\frac{n}{k} - \left\lfloor \frac{n}{k} \right\rfloor \right).$$

Donald recognized 0.422784 as being the first six digits of $1 - \gamma$, where γ
is the Euler-Mascheroni constant, and he intended to prove it in our inaugu-
ral session. Letting "log" denote "natural logarithm", the Euler-Mascheroni
constant is

$$\gamma = \lim_{n \to \infty} \left(\sum_{k=1}^{n} \frac{1}{k} - \log n \right) = 0.5772156\ldots.$$

This constant appears frequently in number theory. It was certainly well
known to Donald, who even detected its presence when it was subtracted
from 1.

About a half dozen of us trooped into the small room that Monday
to hear Donald hold forth. He began enjoying himself, chipping away at
it. Since he was using a technique with which I was vaguely familiar, I was
actually able to follow him. At the thirty-minute mark Donald got stuck. He
picked up the marker and then put it back down again. The cycle repeated
itself several times. We were at a loss to help. He had come into the room
certain that he could negotiate the twists and turns to a proof. In fact,
Donald had not tried to negotiate that path beforehand. He just assumed
that the necessary insights would come to him. At the sixty-minute mark, we
called it a day. As we filed out of the room Donald and I lagged behind. We
watched the rest of his audience move down the hallway. Then he turned to
me and said, "I think it was good for the young people to see the grandmaster
screwing up." That is almost an exact quote, except for one word, which
may be assumed to have been at least one level stronger.

By Thursday I had taken time to go over my notes from the seminar and
believed that I could get past Donald's stumbling block. I went to him and
told him so. His words: "I tried to do it that way and couldn't." He said no
more. I think it was supposed to be a "word to the wise."

Over the weekend, I fashioned a proof based on his attack. When we
went back into the room on Monday, another member of the seminar had
come up with an entirely different approach. Eliana LeVine presented a nice
proof and we all applauded. She was a young mathematician and seemed to
think that we somehow already knew how she was going to do it. We did not.

I followed with my adaptation of the Donald Newman method. Thirty minutes into it, I turned to Donald and said, "I've been a little sloppy, but I think that you can see that this works." Donald's response was, "Yes, despite the sloppitude, it is a proof."

There was silence after that and then Donald spoke again:

"That's fascinating." Pause. "I tried to do that ... and didn't." Pause. "And you did." Pause. "That's fascinating."

A somewhat less sloppy version of my proof is in Appendix E.

I was expecting a Stanford graduate student, Edray Goins, to arrive to begin his second summer internship with me. After Newman's fizzle and my fascinating resuscitation of the approach, I had sent Edray an email message describing the problem. He got back to me in less than twenty-four hours with an entirely different and relatively easy proof. When I handed it to Donald, he read it quickly and then dropped the sheet of paper on his desk, almost in disbelief. He said, "That's the *only* way to prove it." It was from The Book. He started calling Edray "the kid." He was eager to meet the kid. As it turns out he never did. Edray arrived after Donald left for the summer. By the time Donald returned, Edray was back at Stanford. Edray and I wrote a paper[1] with the $1 - \gamma$ result as a starting point. We generalized it and drove it into the ground, nineteen pages!! Below is Edray's proof, which now seems to be the obvious way to go. I've supplied more details than Edray did in his message. We show below that if we pick a function cleverly and evaluate its integral, we get a very neat proof.

$$\text{Let } f(x) = \frac{1}{x} - \left\lfloor \frac{1}{x} \right\rfloor \text{ on } (0, 1].$$

We now collect the pieces to set up the Riemann sum, which turns out to be the average value of the fractional parts of n:

$$f\left(\frac{k}{n}\right) = \frac{n}{k} - \left\lfloor \frac{n}{k} \right\rfloor,$$

$$\frac{1}{n}\sum_{k=1}^{n} f\left(\frac{k}{n}\right) = \frac{1}{n}\sum_{k=1}^{n}\left(\frac{n}{k} - \left\lfloor \frac{n}{k} \right\rfloor\right),$$

$$\int_0^1 \frac{1}{x} - \left\lfloor \frac{1}{x} \right\rfloor \, dx = \lim_{n\to\infty} \frac{1}{n}\sum_{k=1}^{n}\left(\frac{n}{k} - \left\lfloor \frac{n}{k} \right\rfloor\right).$$

Now all we need to do is evaluate the integral.

Let $u = \frac{1}{x}$, which tells us that $\frac{du}{dx} = -\frac{1}{x^2}$.

This allows us to rewrite the integral:

$$\int_0^1 \frac{1}{x} - \left\lfloor \frac{1}{x} \right\rfloor \, dx = -\int_\infty^1 \left(\frac{u - \lfloor u \rfloor}{u^2}\right) du = \int_1^\infty \left(\frac{u - \lfloor u \rfloor}{u^2}\right) du,$$

---

[1] *The Fractional Parts of $\frac{N}{k}$*, M.R. Currie and E.H. Goins, Contemporary Mathematics Volume **275**, pp. 13-31, 2001

$$\int_1^\infty \left(\frac{u - \lfloor u \rfloor}{u^2} \right) du = \lim_{n\to\infty} \sum_{k=1}^{n-1} \int_k^{k+1} \frac{u - k}{u^2} du = \lim_{n\to\infty} \sum_{k=1}^{n-1} \left(\log u + \frac{k}{u} \right) \Bigg|_k^{k+1} .$$

Taking advantage of the fact that many of the terms in the last expression cancel each other, we have things reduced to

$$\lim_{n\to\infty} \left(\log n - \sum_{k=2}^n \frac{1}{k} \right) = \lim_{n\to\infty} \left(1 + \log n - \sum_{k=1}^n \frac{1}{k} \right) = 1 - \gamma.$$

Donald once told me that as a boy, he did not know that you could earn money doing mathematics. His goal was to get a job at a movie theater at the ticket counter, so that he would have long periods of time when he could think just about mathematics.

Not too long after Newman landed in my division Leo Flatto also joined us. Leo had seen the bombs falling on Antwerp at the outset of World War II, when he was ten years old. His family escaped Europe and made it to the United States by way of Casablanca. I seated him in the same cubicle as Newman. They knew each other going back to their graduate days in Cambridge, Massachusetts. Flatto was at MIT. So for about a year I had my two grandfathers at work with me. Jeff Lagarias has put together a hilarious compilation of anecdotes that he calls *The Leo Collection*, anecdotes and stories told by Leo about the trials and tribulations of graduate school. I hope that Jeff puts it out to a wider readership someday. Unfortunately, the sabbatical program eventually ended for Leo and Donald. I held a little going-away party for them at my house. I even performed one of my piano compositions for them. I am not sure that Donald liked it. He said, "It sounds like music, only slower."

In the wake of Donald and Leo's departure Kent Boklan did his first tour in the Agency's Mathematics Development Program and it was as an intern with me. Kent was a proud and newly minted PhD in analytic number theory, which he had earned at the University of Michigan. When Newman left, I asked Kent to take over running the number-theory seminar. He was willing to do this and kicked off his tenure with a two-session talk, in which he painstakingly covered a proof of Brun's Theorem, which states that the sum of the reciprocals of the twin primes is bounded. I believe that he showed that 10 is an upper bound. I don't think that anyone in the room followed much of it after the first fifteen minutes of the first session. At any rate, Kent was not only a proud young man, he was also quite a ham. As he finished the second session and awaited adulation, I walked to the board and announced that I hadn't been able to understand all the details of his very fine proof, but I was inspired by it during the past week to consider some of the methodology. I had not been able to find a specific upper bound, but I had been able to show that the sum must be irrational. Kent knew that I was lying, but was dumbfounded by the audacity and just stood there with his mouth open. I had stolen his thunder. As we were leaving the room he finally found his tongue and demanded that I tell them that it was a joke.

I wouldn't. Later that day one of the seminar participants phoned me to ask if I realized that what I had shown amounted to a proof of the Twin Prime Conjecture. (The sum of only finitely many rational numbers, which the reciprocals of primes are, would be rational.) That's what I was waiting for. I walked over to Kent's desk and said, "They think I've proven the Twin Prime Conjecture." Livid was young Boklan.

Chapter 20

From the Sublime to the Ridiculous

In 1893 James Joseph Sylvester posed the following question. Let $\mathbf{P}$ be a finite set of points in the plane. If every line that contains two of the points in $\mathbf{P}$ also contains a third point in $\mathbf{P}$, must all of the points in $\mathbf{P}$ lie on the same line? This question became known as Sylvester's Problem (not to be confused with Sylvester's Triangle Problem mentioned in the *Euler Line* chapter). Sylvester's question was not answered until 1941 by Eberhard Melchior. The answer is yes.

It is interesting and useful to state the question in an equivalent way, the contrapositive: If we have a finite set $\mathbf{P}$ of points in the plane that do not all lie on the same line, must there exist a line that contains precisely two (no more) of them?

The following elegant proof is due to Leroy Milton Kelly. He apparently produced it in the mid-1940s, but a precise date is not known to the author.

Suppose that the points in $\mathbf{P}$ are not all on the same line and every line that contains two points in $\mathbf{P}$ contains at least three. There are only finitely many lines k containing two points of the set S, since there are only finitely many points in $\mathbf{P}$ and two points determine a line.

Let $\mathbf{K}$ be the set of all lines k that contain two points in $\mathbf{P}$ (and then by assumption at least three). Since both $\mathbf{K}$ and $\mathbf{P}$ are finite sets, we can find a point p in $\mathbf{P}$ and a line k in $\mathbf{K}$ such that the distance between p and k is the minimum *positive* distance for all possible choices. Call this distance d^*. We can do this because there are only finitely many choices and we have assumed that the points in $\mathbf{P}$ do not all lie on the same line. Let p^* and k^* be the point and line that exhibit the minimum distance d^*. Then we have the following picture.

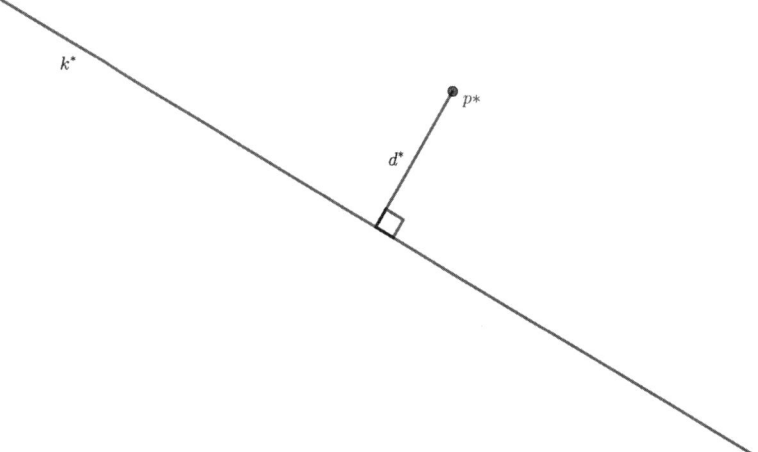

Since k^* is in $\mathbf{K}$ it contains three points in $\mathbf{P}$, and at least two of them must be on the same side of the perpendicular segment from p^* to k^* (include the foot of the perpendicular when we say "same side"). We call these two points a and b and let a be closer to the perpendicular segment than b.

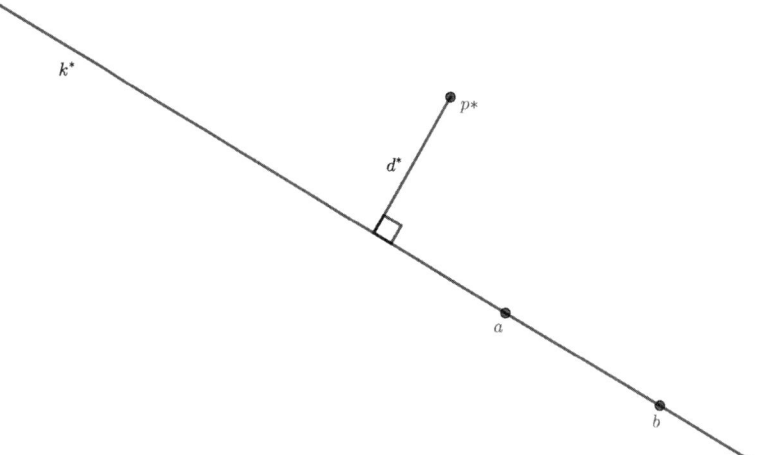

Here is how our assumption of a least positive distance leads to a contradiction. In the following figure we draw the line through p^* and b, call it k^{**}. Then the line k^{**} must be in $\mathbf{K}$, since it contains two points in $\mathbf{P}$. Triangle baQ' is similar to triangle bp^*Q and also smaller. The distance from a to the line k^{**} is d^{**}, which is shorter than the distance d^*, the length of the corresponding side Qp^*, and we have our contradiction.

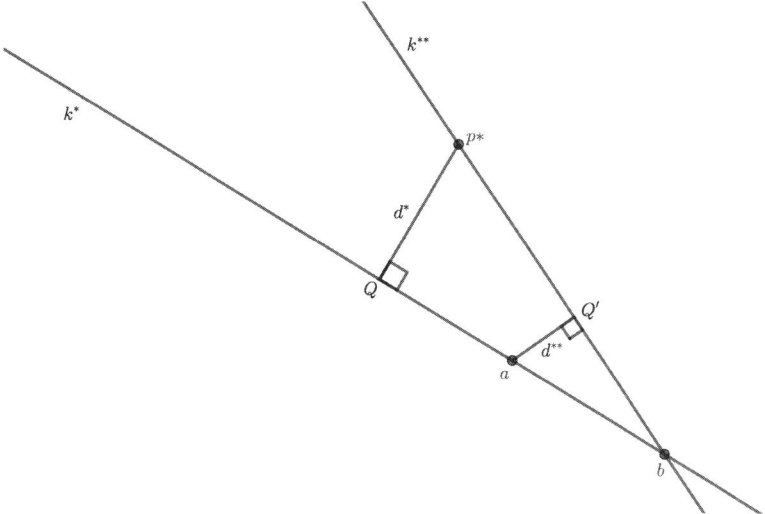

We conclude that if we have a finite set **P** of points that do not all lie on the same line, there must exist a line that contains precisely two of them, which answers Sylvester's question.

* * *

There is no end to novelties. They are around every corner. Here's one now. In the graphic below, $\theta(x)$ is the size of the angle opposite the edge of length b.

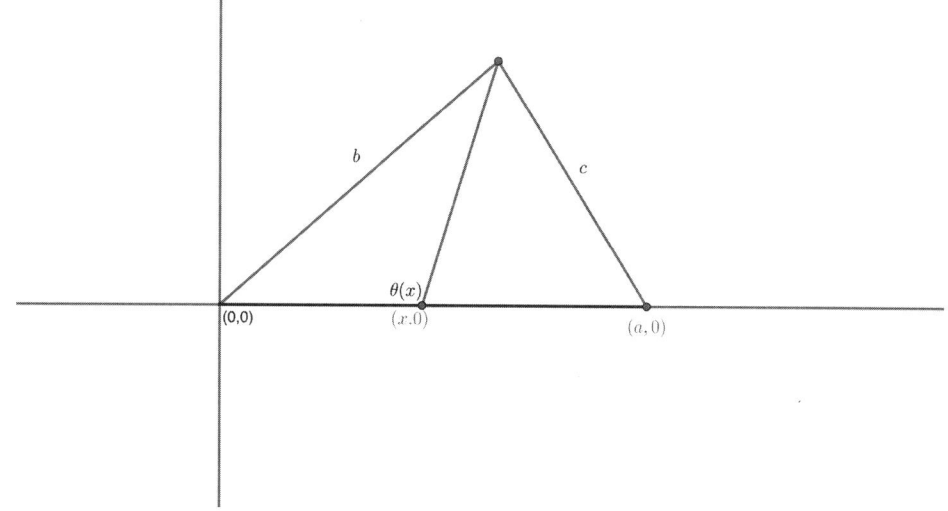

$$\int_0^a \cos\theta\,(x)\,dx =?$$

This is the sort of problem that I am usually willing to tackle. It is simply stated. Let's make a construction. After all, it is geometry. Draw the segment perpendicular to the x-axis between (a_1, a_2) and $(a_1, 0)$ as pictured.

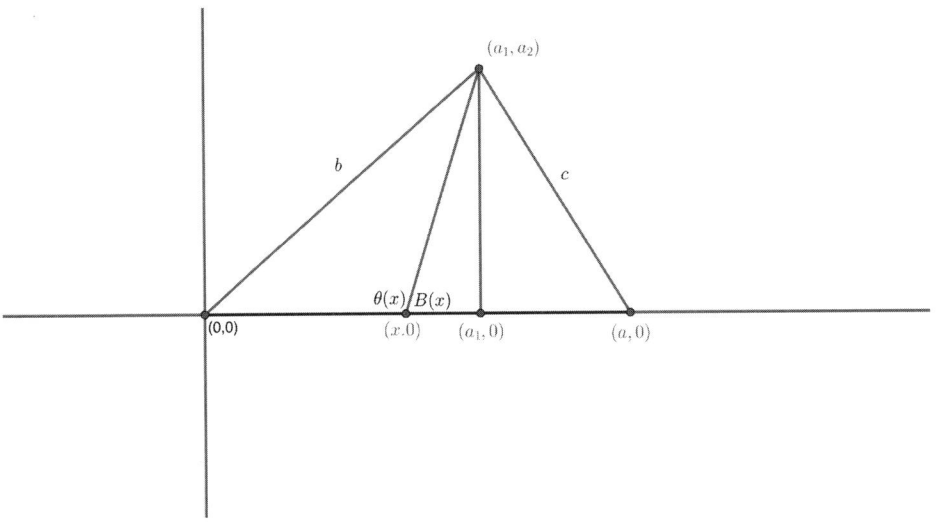

$$\cos \theta \left(x \right) = - \cos B \left(x \right) = \frac{x - a_1}{\sqrt{\left(a_1 - x \right)^2 + a_2^2}}.$$

We have

$$\int_0^a \cos \theta \left(x \right) dx = \int_0^a \frac{x - a_1}{\sqrt{\left(a_1 - x \right)^2 + a_2^2}} dx.$$

A change of variable is helpful here, so we let

$$u = \left(a_1 - x \right)^2 + a_2^2, \; du = 2 \left(x - a_1 \right) dx, \; \text{so} \; \left(x - a_1 \right) dx = \frac{1}{2} du.$$

We adjust the limits of integration to align them with the new variable:

$$x = 0 \Rightarrow u = a_1^2 + a_2^2 = b^2. \; x = a \Rightarrow u = \left(a_1 - a \right)^2 + a_2^2 = c^2.$$

When we rewrite the integral, we have

$$\frac{1}{2} \int_{b^2}^{c^2} \frac{du}{\sqrt{u}} = \sqrt{u} \Big|_{b^2}^{c^2}$$
$$= c - b.$$

This is a gift. Simply formulated. Not too hard to solve. And quite surprising.

$$* * *$$

The *magic square* is a curiosity. Magic squares are arrays of numbers with the number of rows equal to the number of columns and two conditions are satisfied.

(1) Each number in the square occurs only once.

(2) The numbers must be arranged so that the sum of the numbers in each column, each row, and each diagonal are all the same.

Here is the standard example with sum 15:

$$\begin{bmatrix} 2 & 9 & 4 \\ 7 & 5 & 3 \\ 6 & 1 & 8 \end{bmatrix}.$$

A magic square that has all the numbers from 1 to n^2 is called a *normal* magic square.

Here is another magic square. It is not normal:

$$\begin{bmatrix} 5 & 22 & 18 \\ 28 & 15 & 2 \\ 12 & 8 & 25 \end{bmatrix}.$$

Actually, it is not normal in another way.

$$\begin{bmatrix} \text{five} & \text{twenty-two} & \text{eighteen} \\ \text{twenty-eight} & \text{fifteen} & \text{two} \\ \text{twelve} & \text{eight} & \text{twenty-five} \end{bmatrix}.$$

Above we have simply replaced the number with the English word for the number. Now count the number of letters in each word and replace the word with that number to get:

$$\begin{bmatrix} 4 & 9 & 8 \\ 11 & 7 & 3 \\ 6 & 5 & 10 \end{bmatrix}.$$

We have another magic square with sum 21. This borders on madness and offers a nice segue to the promised descent into the ridiculous, which we accomplish in the next segment of this chapter.[1]

* * *

My experience leads me to believe that starting a discussion among mathematicians about anagrams is like introducing a virus into a vulnerable population. I did this once while I was the titular head of an organization and the disease took a long time to cure. If truth be told, I probably suffered from the illness longer than the people in my charge. At some point my wife asked (told) me to stop. I was bringing my illness home with me from work. I was anagramming my colleagues' names and it started spreading to my family members.

An anagram of a person's name is like the message in a fortune cookie. You can always choose to see a little bit of truth in it. This can become something of an issue if you are delivering these fortunes to people you work

---

[1] I thank Boyd Livingston for showing me this magic-magic square.

with every day. I would say to them after I had crafted the anagram: "Don't shoot the messenger." Here is a sampling:

Wart Border

Fat Jerk Sam

Zero Brawn

Cod Row Swarthy

O, I'm a pout

Egoless Rogue

Peach-Lip Homer

I anagrammed **Donald Newman** to

Wan Old Man Ned.

He defended himself with

Land and Women.

Looking around my office one day, I noticed that the previous occupant had left a few things behind, remnants. His last name was Egio. Off into a fantasy I flew and the diagram below was the result.

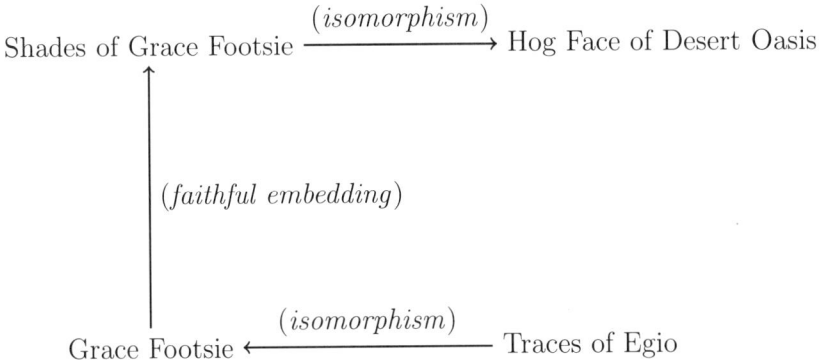

Of course, you could try to make sense of this . . . but you would fail.

I did not spare myself:

Mel Currie

Mice Ruler

Cruel Emir

Melvin Robert Currie

Crime-'n'-Trouble River

Melvin Currie

Vile Miner Cur

Numeric Liver

I think we have hit bottom and found the ridiculous.

Chapter 21

A Few More Words

This has probably gone on long enough for the reader to get my drift. I had a dream as I was completing the last couple of chapters. I was in a class taking a math test. The whole semester I had been under the impression that the instructor was supportive of my efforts to learn a little bit more about mathematics. In the middle of the test I realized that there were not enough pages in the booklet to complete the proofs and computations that I needed to demonstrate my knowledge. Going to the instructor's desk, I asked permission to add some loose sheets so that I could finish. He told me that he would not allow it. Then he said more. I heard the words, "You are not a visionary mathematician." I thought that these words may be true, but that was no reason for not letting me do whatever I could do. At that moment it would have meant completing the test.

I ran track when I was a young man. I called myself a sprinter. I did get to stand on the same track with some of the great runners in the country and try to acquit myself as well as I could. In college it was Ivory Crockett, John Carlos, Erv Hall, Larry James, Frank Shorter, ... In high school it was Orin Richburg. No one will remember me, but I was picking up my feet and putting them down.

In the chapter on Sperner's Lemma I reflected on how often it occurs that properties of the interior of a geometric object can be discerned by examining its surface. Generalizing, maybe some of us can make a contribution just by scratching the surface of mathematics. That infinite surface .... You can slide along it and feel the heartbeat.

I remember meeting David Blackwell at the Joint Mathematics Meetings one year in San Francisco. My Blackwell family roots were in Virginia, as were his. We never figured out if we really went back to a common ancestor, some poor soul who toiled on a Virginia farm in the 1700s, but we were both mathematicians. J. Ernest Wilkins was heralded as the "Negro genius" in 1942, when as a nineteen-year-old he completed his PhD at the University of Chicago. Blackwell did not complete the PhD until he was at the rather advanced age of twenty-two. Maybe that is how he avoided the Negro-genius moniker. In 1996, when I was giving a talk at a conference, J. Ernest Wilkins stole my thunder by blurting out, "This must be where the di-gamma function comes in," spilling the mathematical beans, just as I was poised to spring

it on the audience. I forgive him. He couldn't help it. I never should have
mentioned the di-gamma function explicitly in the abstract. As fate would
have it, I was honored in 1999 to give the David Blackwell Lecture at the
Mathematical Association of America's Mathfest and in 2009 the J. Ernest
Wilkins Lecture at the National Association of Mathematicians' Mathfest.
The title of the talk in 2009 was *Trivialities I Have Known*, and it gave rise
to this book.

I have been able to lift my feet up and put them down on the same track
as J. Ernest Wilkins, Abraham Robinson, Walter Feit, Gustav Hedlund,
Donald J. Newman, Yitang Zhang.... And there was David Blackwell, who
always said he was not trying to do research as much as he was trying to
understand. Somehow, he managed to produce at least sixty-five doctoral
students at Berkeley.

I never supervised a doctoral dissertation during my brief seven-year
academic career. I did, however, have the great pleasure of supporting a
large number of interns during my twenty-five years at the National Security
Agency. The projects ranged from cryptanalysis to quantum computing algorithms, from cryptographic design to number theory. I mentioned earlier
my last intern, who in her junior year in high school worked with me on
a project that used Hidden Markov Modeling to analyze the biogeographical contributions to my X chromosome. Mathematics took me to Canada,
England, and Australia to confer with colleagues at NSA's sister agencies in
those countries. And mathematics took me to the Salish-Kootenai Reservation in Montana and the Navajo Reservation in the Southwest, where young
people were willing to listen to me talk about the subject.

When I was a professor at the University of Richmond, a mother had
the courage to sit in on my calculus class with her daughter on Parents Day.
She was the only parent there. Afterwards she approached me, probably
embarrassing her daughter tremendously, and told me how much she appre-
ciated the lecture. She engaged me in conversation and continued to laud
me, superlative after superlative. I drifted to my office and she followed, her
daughter walking a bit behind the two of us. When she got to my office, she
seemed to run out of things to say. She stopped for a moment and then said
in all seriousness, "You are so good you could have been an engineer." I was
speechless.

Taken from *A Mathematician's Apology* by G. H. Hardy[1]

> I was at my best a little past forty, when I was a professor at
> Oxford. Since then I have suffered from that steady deterioration
> which is the common fate of elderly men and particularly of elderly
> mathematicians. A mathematician may still be competent enough
> at sixty, but it is useless to expect him to have original ideas
> ...If I had been offered a life neither better nor worse when I was

---

[1] *A Mathematician's Apology*, G. Hardy and C. Snow, Cambridge University Press,
pp. 48–50. doi:10.1017/CBO9781139644112.003

twenty, I would have accepted without hesitation. It seems absurd to suppose that I could have "done better."

My choice was right, then, if what I wanted was a reasonable, comfortable and happy life. But solicitors and stockbrokers and bookmakers often lead comfortable and happy lives, and it is very difficult to see how the world is richer for their existence. Is there any sense in which I can claim that my life has been less futile than theirs? It seems to me again that there is only one possible answer: yes, perhaps, but, if so, for one reason only: I have never done anything "useful." No discovery of mine has made, or is likely to make, directly or indirectly, for good or ill, the least difference to the amenity of the world. I have helped to train other mathematicians, but mathematicians of the same kind as myself, and their work has been, so far at any rate as I have helped them to it, as useless as my own. Judged by all practical standards, the value of my mathematical life is nil; and outside mathematics it is trivial anyhow. I have just one chance of escaping a verdict of complete triviality, that I may be judged to have created something worth creating. And that I have created is undeniable: the question is about its value. The case for my life, then, or for that of any one else who has been a mathematician in the same sense which I have been one, is this: that I have added something to knowledge, and helped others to add more; and that these somethings have a value which differs in degree only, and not in kind, from that of the creations of the great mathematicians, or of any of the other artists, great or small, who have left some kind of memorial behind them.

I am not in complete agreement with everything that Hardy asserts in this passage. Creativity is not just for the young, but for many people it is during their youth that they display whatever originality they have to offer. We see this in many areas of human endeavor. As the composer ages, we can frequently hear the signature of his youth in his later work. On the matter of futility, the brain is part of the universe. Unleashing it to explore mathematics should shed light on the universe in which it resides, whether that is the goal or not. However, the spirit of the excerpt from his *Apology* resonates with me.

> I sent my Soul through the Invisible,
> Some letter of that After-life to spell:
> And by and by my Soul return'd to me,
> And answer'd "I Myself am Heav'n and Hell:"
>
> *Omar Khayyám*

Angela and Rudy

This book is dedicated to the memory of Rudy Horne and Angela Grant, two people I got to know as young mathematicians. Angela was one of my summer interns just before she began graduate school and we remained friends until her death at age 36 in 2010. When we talked, I would frequently make her laugh, although often I wasn't joking. I also got to know Rudy Horne when he was a graduate student. We were both in regular attendance at Bill Massey's annual Conference for African American Researchers in the Mathematical Sciences (CAARMS). He would say that I was his hero, which always embarrassed me. Ironically, now that he has passed away, I wish I had found the courage to ask him why. Rudy was a professor at Morehouse when he died in 2017 at age 49. In 2016 he had served as the mathematics consultant for the film *Hidden Figures*.

Photos and Pictures

Chapter 2. Primes

Yitang Zhang[2]

Chapter 5. Some Things Add Up. Some Don't.

In a couple of years Petra would tackle Zeno's Paradox. Her puppy was enough in this photo.

---

[2]Wikimedia Commons

Chapter 9. Impossibilities

Évariste Galois[3]

Neils Henrik Abel[4]

---

[3]Wikimedia Commons
[4]Wikimedia Commons

Chapter 7. Plus or Minus

1966 Senior Year, Calculus with Mr. Levy

Standing next to Alaska Residence, Summer 1970

Gulf Oil 1971

Chapter 10. Magnitudes of Infinity

Georg Cantor[5]

Abraham Robinson[6]

Düsseldorf, die Kaufmännische Schule IV

Chapter 12. Consider the Sequence

Paul Erdős[7]

Chapter 17. Bulgarian Solitaire

At the State Department: Banquet for American Mathematical Olympiad winners (left to right, Shirley Currie, Mel Currie, Bob Cordwell, Bill Cordwell, Rosemary Cordwell), 2005

---

[7]Photo of Paul Erdős taken in Poznan, Poland by George Csicsery in 1989 @ZalaFilms. All rights reserved.

Chapter 19. Fascinating

D. J. Newman-Currie-Leo Flatto (left to right) in the Currie parlor, 1999

Chapter 21. A Few More Words

(l-r) Nathaniel Dean, David Blackwell, Currie at the Joint Mathematics
Meetings in San Francisco, 1995

The annual Conference for African Americans in the Mathematical Sciences in 2000 A few of the attendees in front of the author's home - from the back: Earl Barnes (Georgia Tech), William Massey (Princeton), Carl Graham (Ècole Polytechnique), Currie (National Security Agency), Ahmad Ridley (Graduate Student, University of Maryland), Edray Goins (Post-doc, Institute for Advanced Study), Kathryn Lewis (Morehead State University). To the right: Scott Williams (University at Buffalo, SUNY)

Currie and Rudy Horne (1968-2017)

A Few of the Author's Interns

Patricia   Greene   Xiaolu Guo (last intern)
(first intern)

With Edray Goins in Currie Dining Room   Angela Grant
                                                   1973-2010

In the West

The Math Camp held at Diné College on the Navajo Reservation. Back Row: Henry Fowler (Diné College), David Auckly (Kansas State University), Currie, Matthias Kawski (Arizona State University), Tatiana Shubin (San Jose State University). Front Row: Albert Haskie, Charmayne Seaton, Natanii Yazzie, 2016

Appendix A

Notation, etc.

In this appendix we discuss a few topics that might not be treated in a high school algebra course.

Modular Arithmetic

We say that an integer m is equal to j *modulo* an integer N if for some integer k,

$$m = kN + j.$$

We write this with the following notation:

$$m = j \text{ modulo } N.$$

The word **modulo** is usually abbreviated to "mod."

So m and n equal **mod** N if their difference is divisible by N. We write

$$m = n \quad \text{mod } N.$$

In particular, writing "$m = 0 \mod N$" is equivalent to writing "m is divisible by N."

Vectors

First, we tackle the basic vector concept. We identify each vector with a point (a_1, a_2) in the Cartesian plane. The vector will have two features. The first is its length. For any point (a_1, a_2), the corresponding vector will have length equal to the distance from the point to the origin. The second feature will be direction. Perhaps the most straightforward way to capture direction is to define it to be the angle that the line segment from the origin to the point forms with the x-axis. It is measured counterclockwise from the x-axis. It is a convention to picture a vector as an arrow of length $\sqrt{a_1^2 + a_2^2}$ drawn from the origin to the point (a_1, a_2), with which the vector is identified.

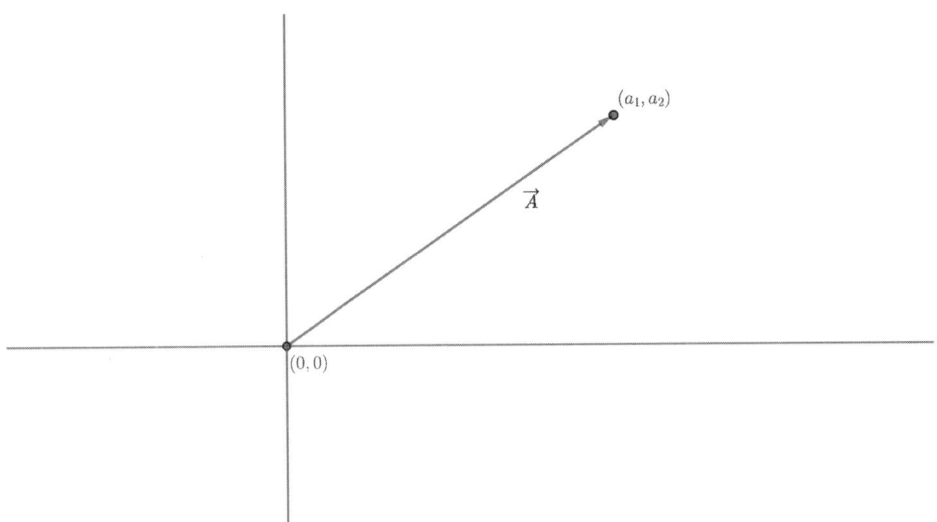

We define arithmetic operations on vectors as follows, where c is a real number:

$$c\vec{A} = (ca_1, ca_2),$$
$$\vec{A} + \vec{B} = (a_1 + b_1, a_2 + b_2).$$

If $c > 0$, $c\vec{A}$ and $\vec{A}$ have the same direction. If $c < 0$, $c\vec{A}$ points in the opposite direction from $\vec{A}$. The midpoint of the line segment from the tip of $\vec{A}$ to the tip of $\vec{B}$ is one half the sum of the two vectors:

$$\frac{1}{2}\left(\vec{A} + \vec{B}\right) = \left(\frac{a_1 + b_1}{2}, \frac{a_2 + b_2}{2}\right) = \vec{B} + \frac{1}{2}\left(\vec{A} - \vec{B}\right).$$

The rightmost expression allows us to visualize finding the midpoint of the segment by placing the tail of vector $\frac{1}{2}(\vec{A} - \vec{B})$ at the tip of $\vec{B}$. It is common practice to move vectors in this way to emphasize certain relationships. Addition is more often than not displayed as a tip-to-tail connection. The reader can quickly verify that $\frac{1}{2}(\vec{A} + \vec{B})$ is indeed the midpoint of the line segment connecting (b_1, b_2) to (a_1, a_2) by first checking to see that it is on the line connecting the points and then that it is equidistant from them.

The second thing that we want to observe is that if

$$\vec{C} - \vec{D} = \vec{A} - \vec{B},$$

the line segment that connects the tip of $\vec{C}$ to the tip of $\vec{D}$ must be parallel to, and have the same length as, the line segment that connects the tip of $\vec{A}$ to the tip of $\vec{B}$.

Set Notation

If A and B are collections (we will say *sets*) of objects (*elements*), then the set C that contains precisely those elements that are in A **or in** B is called the *union* of A and B. We write:

$$C = A \cup B.$$

The set D that contains precisely those elements that are in both A *and* B is called the *intersection* of A and B. We write:

$$D = A \cap B.$$

If A is contained in B, we write

$$A \subseteq B.$$

We let the empty set (no elements) be represented by the symbol $\emptyset$. We have

$$A \cup A = A,$$
$$A \cap A = A,$$
$$\text{and } A \cap (B \cup C) = (A \cap B) \cup (A \cap C).$$

This last line looks very much like the distributive law that we have for real numbers, $a \times (b + c) = a \times b + a \times c$. We actually have this law for English prepositions too, but this is frequently forgotten by some native speakers.

for you and for me
$\neq$
for you and I

Man against distributive lawlessness

Summations

We use the summation notation as a compact way of describing sums. If we have an expression $E(x)$ that we are evaluating at $k = 1, 2, 3, 4,$ and 5, then adding up the five numbers that are produced, we could write:

$$E(1) + E(2) + E(3) + E(4) + E(5).$$

Using the summation notation, we can write this in a more economical way as

$$\sum_{k=1}^{5} E(k).$$

This is obviously very useful notation if we have a great many terms, perhaps even infinitely many. If we want to add over just those numbers that satisfy a certain criterion C, we can write

$$\sum_{C} E(k),$$

where enough information is provided in criterion C to allow us to know which integers are in play.

Limits

Suppose we want to assess the behavior of an expression $T(n)$ as n gets arbitrarily large. We are looking to see if we can establish the behavior of $T(n)$ as n goes to infinity. We write

$$\lim_{n\to\infty} T(n).$$

For instance,

$$\lim_{n\to\infty} \frac{n+1}{n} = \lim_{n\to\infty} \left(1 + \frac{1}{n}\right) = 1.$$

The second equality follows from the fact that $1/n$ gets arbitrarily close to zero as n gets large. That is, we can make $(1 + 1/n)$ as close to 1 as we would like just by assuring that we have taken n to be sufficiently large.

We define

$$\sum_{k=1}^{\infty} E(k)$$

to be

$$\lim_{n\to\infty} \sum_{k=1}^{n} E(k).$$

It is generally true that when we tackle the limiting process we can find that the limit does not exist. That discussion would take us too far afield in this review of notation.

Proof by Induction

Suppose that we are trying to prove that some claim, $P(n)$, holds for all n greater than or equal to j. Proof by induction is sometimes a useful tool. We outline the principle.

Basis for induction: Show that $P(j)$ holds. In many cases $j = 0$ or $j = 1$.

Inductive Step: Show that if $P(k)$ is true, then $P(k + 1)$ is also true. Since all the integers greater than j are greater than their predecessor by 1, $P(n)$ must hold for all n greater than or equal to j.

Let's try it out. We'll prove that

$$10^{2n+1} + 1$$

is always divisible by 11, for n greater than or equal to zero.

Let's establish the basis at $j = 0$:

$$10^{2 \cdot 0 + 1} + 1 = 10 + 1 = 11.$$

Assume that

$$10^{2k+1} + 1$$

is divisible by 11. Then

$$10^{2(k+1)+1} + 1 = 10^2 \left(10^{2k+1}\right) + 1 = 10^2 \left(10^{2k+1} + 1\right) - \left(10^2 - 1\right).$$

The first term in the rightmost expression is divisible by 11 by the inductive assumption and the second term, -99, is also divisible by 11, so the whole expression must be. This completes the inductive step and establishes the truth of our claim.

Appendix B

Mysterious

Here we derive the sequence that was presented in the *Mysterious Pattern* chapter. As a reference, the first few terms of the sequence are

$$a_0 = 2,$$

$$a_1 = 2\sqrt{2},$$

$$a_2 = 2^2\sqrt{2 - \sqrt{2}},$$

$$a_3 = 2^3\sqrt{2 - \sqrt{2 + \sqrt{2}}},$$

$$a_4 = 2^4\sqrt{2 - \sqrt{2 + \sqrt{2 + \sqrt{2}}}},$$

$$a_5 = 2^5\sqrt{2 - \sqrt{2 + \sqrt{2 + \sqrt{2 + \sqrt{2}}}}},$$

$$a_6 = 2^6\sqrt{2 - \sqrt{2 + \sqrt{2 + \sqrt{2 + \sqrt{2 + \sqrt{2}}}}}}.$$

Pictured below is an octagon inscribed in a circle of radius 1.

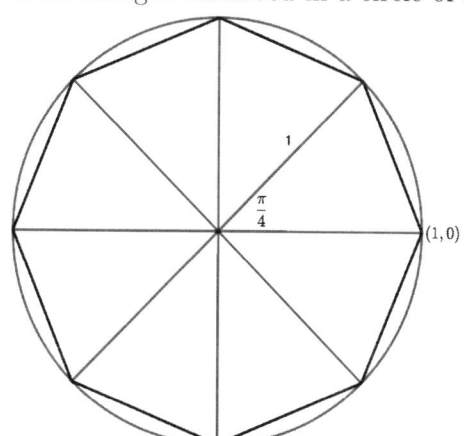

The area of each of the eight congruent triangles is $\frac{1}{2} \times \frac{\sqrt{2}}{2}$, so the area of the octagon is eight times that value, $2\sqrt{2}$.

It is useful to define the sequence formally:

$$\text{Let } b_0 = -2 \text{ and } b_n = \sqrt{2 + b_{n-1}}, \text{ for } n \geq 1.$$

$$\text{Then } a_n = 2^n \sqrt{2 - b_n}, \text{ for } n \geq 0.$$

We have seen that this formula does represent the area of the inscribed octagon for $n = 1$. The case $n = 0$ clearly represents the inscribed square. If we can show that the area of the triangle associated with the 2^{n+2}- edge polygon gives rise to a_n, we're done. The associated triangle is pictured below.

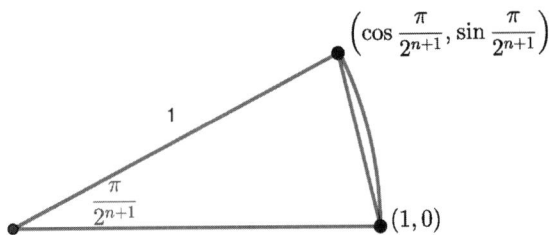

The area of the polygon is 2^{n+2} times the area of the triangle and the area of the triangle is

$$\frac{1}{2} \sin \frac{\pi}{2^{n+1}}.$$

Now we develop a formula for $\sin \frac{\pi}{2^{n+1}}$.

By the half-angle formulae, we have

$$\sin \frac{\pi}{2^{n+1}} = \sqrt{\frac{1 - \cos \frac{\pi}{2^n}}{2}} \quad \text{and}$$

$$\cos \frac{\pi}{2^n} = \pm \sqrt{\frac{1 + \cos \frac{\pi}{2^{n-1}}}{2}}. \quad (\text{For } n = 0 \text{ we need the minus sign.})$$

We will show by induction that

$$\cos \frac{\pi}{2^n} = \frac{1}{2} b_n.$$

For $n = 0$, $\cos \pi = -1 = \frac{1}{2} b_0$, which establishes the basis for induction.

Assume that for $n = k$

$$\cos \frac{\pi}{2^k} = \frac{1}{2} b_k.$$

For $n = k + 1$

$$\cos \frac{\pi}{2^{k+1}} = \sqrt{\frac{1 + \cos \frac{\pi}{2^k}}{2}} = \sqrt{\frac{1 + \frac{1}{2} b_k}{2}} = \sqrt{\frac{2 + b_k}{4}} = \frac{1}{2} b_{k+1}.$$

This completes the induction.

To wrap things up we write

$$\sin \frac{\pi}{2^{n+1}} = \sqrt{\frac{1 - \cos \frac{\pi}{2^n}}{2}} = \sqrt{\frac{1 - \frac{1}{2} b_n}{2}} = \frac{1}{2} \sqrt{2 - b_n}.$$

To get the area of the triangle we multiply the last expression by $\frac{1}{2}$ and have

$$\frac{1}{4} \sqrt{2 - b_n}.$$

Finally, we remember that to get the area A of the polygon we have to multiply the area of the triangle by 2^{n+2}. We then get

$$A = 2^n \sqrt{2 - b_n}.$$

We're done.

Appendix C

Impossibilities

Showing the Wonders of Linear Algebra

Let $(0,0)$, (x_1, y_1), (x_2, y_2), (x_3, y_3) be four points in the plane such that the distance between any pair of them is an odd integer. Let matrix A be defined as follows:

$$A = \begin{bmatrix} a_{11} & a_{12} & a_{13} \\ a_{21} & a_{22} & a_{23} \\ a_{31} & a_{32} & a_{33} \end{bmatrix} = \begin{bmatrix} x_1 & y_1 \\ x_2 & y_2 \\ x_3 & y_3 \end{bmatrix} \begin{bmatrix} x_1 & x_2 & x_3 \\ y_1 & y_2 & y_3 \end{bmatrix}.$$

Given that the rank of A is less than or equal to 2, it follows that $det(A) = 0$.

Note that $a_{jj} = x_j^2 + y_j^2$.

Since a_{jj} is the square of an odd integer, the distance from (x_j, y_j) to $(0,0)$, it must equal $1 \mod 8$. Further,

$$\text{for } i \neq j, 2a_{ij} = x_i^2 + y_i^2 + x_j^2 + y_j^2 - \left[(x_i - x_j)^2 + (y_i - y_j)^2 \right].$$

We conclude that $2a_{ij}$ is also equal to $1 \mod 8$, when i is not equal to j. We then see that the matrix $2A$ is entry by entry congruent mod 8 to

$$\begin{bmatrix} 2 & 1 & 1 \\ 1 & 2 & 1 \\ 1 & 1 & 2 \end{bmatrix}.$$

The determinant of this matrix is 4, not 0, a contradiction.

Appendix D

Magnitudes

It is not difficult to show that the set of subsets of the integers has greater magnitude than the set of integers. First, we show that there is a 1-1 correspondence between the real numbers in the interval $(0,1)$ and all positive real numbers. We recall that the tangent function establishes a 1-1 correspondence between the numbers in the interval $(0, \frac{\pi}{2})$ and the positive real numbers. Multiplying each number in the interval $(0,1)$ by $\frac{\pi}{2}$ establishes a 1-1 correspondence between the values in the interval $(0,1)$ and the values in the interval $(0, \frac{\pi}{2})$. We then see that the function

$$f(x) = \tan(\tfrac{\pi}{2}x),$$

for x in $(0,1)$, establishes a 1-1 correspondence between the real numbers in the interval $(0,1)$ and the positive real numbers. Now, each number in the interval $(0,1)$ has a unique binary expansion with infinitely many ones. Each infinite binary expansion can be paired with a unique infinite subset of the positive integers as follows. For a given infinite subset, S, of positive integers, we choose the number r in $(0,1)$ that has 1 in its mth binary place, if and only if m is in S. Let H be the function that assigns each *proper* infinite subset (exclude the whole set) of the positive integers to a number in $(0,1)$ as we have just described. For instance, the subset of odd integers is assigned to $0.1010101\ldots$ by H.

Then $f(H(x))$ is a 1-1 correspondence between the infinite subsets of the positive integers and the set of positive real numbers. We conclude that there are uncountably many infinite subsets of the positive integers, so the set of all subsets of the integers must have cardinality greater than the set of positive integers, which is, of course, countable.

Appendix E

Fascinating

In 1997 Edray Goins and I put out a paper that frolicked through various takes on the asymptotic behavior of the average value of the fractional parts of n/k, where k runs over all positive integers less than or equal to n. In the appendix of that paper I provided an alternate proof to Goins's far simpler and more elegant one. The Goins proof appeared in the *Fascinating* chapter. Recently, I took a look at my alternate proof in the paper and noticed that the upper index in one of the sums was off by 1. On top of this, there are the two figures accompanying the proof, the utility of which is now lost to me. I also introduced sloppiness (actually incorrectness) by waving my hands in the general case, when n is not a square. The revision below is an attempt to address these issues. I do view this appendix as something that I have added for the sake of completeness, since it is the central piece in the Newman story. It will undoubtedly be too advanced for most readers, but it probably does not hurt to look at it. For some readers it will be standard fare in the future.

Here is the definition of the Euler-Mascheroni constant, γ:

$$(1) \qquad \gamma = \lim_{n\to\infty} \sum_{k=1}^{n} \frac{1}{k} - \log n.$$

Let $\lfloor r \rfloor$ denote "the greatest integer less than or equal to r." The "fractional part of r," $\{r\}$, is defined to be $r - \lfloor r \rfloor$. We want to show that

$$(2) \qquad \lim_{n\to\infty} \frac{1}{n} \sum_{k=1}^{n} \left\{ \frac{n}{k} \right\} = 1 - \gamma.$$

It will be useful to establish that

$$(*) \qquad \left\lfloor \frac{n}{b+1} \right\rfloor + 1 \le k \le \left\lfloor \frac{n}{b} \right\rfloor, \ b \text{ and } n \text{ positive, implies that } b \le \frac{n}{k} < b+1.$$

We have:

$$\frac{1}{\left\lfloor \frac{n}{b} \right\rfloor} \le \frac{1}{k} \le \frac{1}{\left\lfloor \frac{n}{b+1} \right\rfloor + 1}.$$

So

$$\frac{n}{\left\lfloor \frac{n}{b} \right\rfloor} \le \frac{n}{k} \le \frac{n}{\left\lfloor \frac{n}{b+1} \right\rfloor + 1}.$$

Looking back at what we were trying to prove, we see that we are done, because

$$\frac{n}{\left\lfloor \frac{n}{b} \right\rfloor} \ge b, \text{ since } \left\lfloor \frac{n}{b} \right\rfloor \le \frac{n}{b}, \text{ and } \frac{n}{\left\lfloor \frac{n}{b+1} \right\rfloor + 1} < b+1, \text{ since } \left\lfloor \frac{n}{b+1} \right\rfloor + 1 > \frac{n}{b+1}.$$

Using (*), we rewrite the sum in (2) as follows:

$$\sum_{k=1}^{n}\left\{\frac{n}{k}\right\} = \sum_{k=\left\lfloor \frac{n}{2} \right\rfloor+1}^{n}\left(\frac{n}{k}-1\right) + \sum_{k=\left\lfloor \frac{n}{3} \right\rfloor+1}^{\left\lfloor \frac{n}{2} \right\rfloor}\left(\frac{n}{k}-2\right)$$

$$+\cdots+ \sum_{k=\left\lfloor \frac{n}{\lfloor\sqrt{n}\rfloor} \right\rfloor+1}^{\left\lfloor \frac{n}{\lfloor\sqrt{n}\rfloor-1} \right\rfloor}\left(\frac{n}{k}-(\lfloor\sqrt{n}\rfloor-1)\right) + \sum_{k=1}^{\left\lfloor \frac{n}{\lfloor\sqrt{n}\rfloor} \right\rfloor}\left\{\frac{n}{k}\right\}$$

(3)
$$=\sum_{k=1}^{n}\frac{n}{k} + \left(n-\left\lfloor \frac{n}{2} \right\rfloor\right)(-1) + \left(\left\lfloor \frac{n}{2} \right\rfloor - \left\lfloor \frac{n}{3} \right\rfloor\right)(-2)$$

$$+\cdots+ \left(\left\lfloor \frac{n}{\lfloor\sqrt{n}\rfloor-1} \right\rfloor - \left\lfloor \frac{n}{\lfloor\sqrt{n}\rfloor} \right\rfloor\right)(-(\lfloor\sqrt{n}\rfloor-1))$$

(4)
$$-\left\lfloor \frac{n}{1} \right\rfloor - \left\lfloor \frac{n}{2} \right\rfloor - \cdots - \left\lfloor \frac{n}{\left\lfloor \frac{n}{\lfloor\sqrt{n}\rfloor} \right\rfloor} \right\rfloor.$$

Notice that there is considerable telescoping in (3). After simplifying, that line becomes:

(3')
$$\sum_{k=1}^{n}\frac{n}{k} - \left(n+\left\lfloor \frac{n}{2} \right\rfloor + \left\lfloor \frac{n}{3} \right\rfloor + \cdots + \left\lfloor \frac{n}{\lfloor\sqrt{n}\rfloor-1} \right\rfloor\right) + \left\lfloor \frac{n}{\lfloor\sqrt{n}\rfloor} \right\rfloor(\lfloor\sqrt{n}\rfloor-1).$$

We'll remove this restriction later, but for now we consider only the case when n is of the form $a^2 + b$, where $0 \le b < a$, a and b integers. Under this assumption, we have

$$\lfloor\sqrt{n}\rfloor - 1 = a - 1 = \left\lfloor \frac{n}{\lfloor\sqrt{n}\rfloor} \right\rfloor - 1.$$

That allows us to consolidate (3') and (4) to write:

$$(5) \quad \sum_{k=1}^{n} \left\{ \frac{n}{k} \right\} = \sum_{k=1}^{n} \frac{n}{k} - 2 \sum_{k=1}^{\lfloor \sqrt{n} \rfloor - 1} \left\lfloor \frac{n}{k} \right\rfloor + \left\lfloor \frac{n}{\lfloor \sqrt{n} \rfloor} \right\rfloor (\lfloor \sqrt{n} \rfloor - 1) - \left\lfloor \frac{n}{\left\lfloor \frac{n}{\lfloor \sqrt{n} \rfloor} \right\rfloor} \right\rfloor.$$

Observe that $\displaystyle -2 \sum_{k=1}^{\lfloor \sqrt{n} \rfloor - 1} \left\lfloor \frac{n}{k} \right\rfloor = -2n \sum_{k=1}^{\lfloor \sqrt{n} \rfloor - 1} \frac{1}{k} + \mathcal{O}\left(\lfloor \sqrt{n} \rfloor - 1 \right)$ and by (1)

$$(6) \qquad\qquad = -2n \left(\log \left(\lfloor \sqrt{n} \rfloor - 1 \right) + \gamma + \epsilon_n \right) + \mathcal{O}\left(\lfloor \sqrt{n} \rfloor - 1 \right).$$

Note that ϵ_n goes to zero as n goes to infinity. We can use (6) to rewrite the second summation on the right-hand side of the equation on line (5). After doing this, and then multiplying both sides of the equation by $\frac{1}{n}$, we again use big-O notation to tidy things up:

$$(7) \quad \frac{1}{n} \sum_{k=1}^{n} \left\{ \frac{n}{k} \right\} = \sum_{k=1}^{n} \frac{1}{k} - 2 \left(\log(\lfloor \sqrt{n} \rfloor - 1) + \gamma \right) + \mathcal{O}\left(\frac{1}{\sqrt{n}} \right)$$

$$+ \frac{1}{n} \left\lfloor \frac{n}{\lfloor \sqrt{n} \rfloor} \right\rfloor (\lfloor \sqrt{n} \rfloor - 1)$$

$$= \sum_{k=1}^{n} \frac{1}{k} - 2 \log(\lfloor \sqrt{n} \rfloor - 1) - \gamma + \frac{1}{n} \left\lfloor \frac{n}{\lfloor \sqrt{n} \rfloor} \right\rfloor (\lfloor \sqrt{n} \rfloor - 1)$$

$$- \gamma + \mathcal{O}\left(\frac{1}{\sqrt{n}} \right) \to 1 - \gamma,$$

since $\sum_{k=1}^{n} \frac{1}{k} - \log n - \gamma \to 0$ implies that $\sum_{k=1}^{n} \frac{1}{k} - 2 \log \left(\lfloor \sqrt{n} \rfloor - 1 \right) - \gamma = \sum_{k=1}^{n} \frac{1}{k} - \log \left(\lfloor \sqrt{n} \rfloor - 1 \right)^2 - \gamma$ does too.

We now address what happens when n is of the form $a^2 + b, a \le b \le 2a$. If $a \le b < 2a$, then

$$\lfloor \sqrt{n} \rfloor - 1 = a - 1 = \left\lfloor \frac{n}{\lfloor \sqrt{n} \rfloor} \right\rfloor - 2.$$

If $b = 2a$,

$$\lfloor \sqrt{n} \rfloor - 1 = a - 1 = \left\lfloor \frac{n}{\lfloor \sqrt{n} \rfloor} \right\rfloor - 3.$$

In the first case line (5) would be

$$\sum_{k=1}^{n} \left\{ \frac{n}{k} \right\} = \sum_{k=1}^{n} \frac{n}{k} - 2 \sum_{k=1}^{\lfloor \sqrt{n} \rfloor - 1} \left\lfloor \frac{n}{k} \right\rfloor + \left\lfloor \frac{n}{\lfloor \sqrt{n} \rfloor} \right\rfloor (\lfloor \sqrt{n} \rfloor - 1)$$

$$- \left\lfloor \frac{n}{\left\lfloor \frac{n}{\lfloor \sqrt{n} \rfloor} \right\rfloor - 1} \right\rfloor - \left\lfloor \frac{n}{\left\lfloor \frac{n}{\lfloor \sqrt{n} \rfloor} \right\rfloor} \right\rfloor.$$

In the second case line (5) becomes

$$\sum_{k=1}^{n}\left\{\frac{n}{k}\right\} = \sum_{k=1}^{n}\frac{n}{k} - 2\sum_{k=1}^{\lfloor\sqrt{n}\rfloor-1}\left\lfloor\frac{n}{k}\right\rfloor + \left\lfloor\frac{n}{\lfloor\sqrt{n}\rfloor}\right\rfloor\left(\lfloor\sqrt{n}\rfloor - 1\right)$$

$$- \left\lfloor\frac{n}{\left\lfloor\frac{n}{\lfloor\sqrt{n}\rfloor}\right\rfloor - 2}\right\rfloor - \left\lfloor\frac{n}{\left\lfloor\frac{n}{\lfloor\sqrt{n}\rfloor}\right\rfloor - 1}\right\rfloor - \left\lfloor\frac{n}{\left\lfloor\frac{n}{\lfloor\sqrt{n}\rfloor}\right\rfloor}\right\rfloor.$$

In the first case we have introduced one additional term, which is $\mathcal{O}(\sqrt{n})$, and in the second case we have introduced two. The reader now sees that, in either case, proceeding as we did earlier from line (5) leaves line (7) unaltered. This completes the proof, but the truth of this fact remains something of a wonder:

$$\lim_{n\to\infty}\frac{1}{n}\sum_{k=1}^{n}\left\{\frac{n}{k}\right\} = 1 - \gamma = 0.4227843351\ldots.$$